ALGEBRA THROUGH APPLICATIONS

with Probability and Statistics

Part I

Student Text

First-Year Algebra via Applications Development Project

Zalman Usiskin, Director

Supported by
National Science Foundation Grant No. SED74-18948

Published by the National Council of Teachers of Mathematics
1906 Association Drive, Reston, Virginia 22091
1979

Any opinions, findings, conclusions, or recommendations
expressed or implied herein are those of the author and
do not necessarily reflect the views of the National Science
Foundation or the University of Chicago.

Comments or questions about this work are invited from any
and all interested parties. Address all correspondence to
Zalman Usiskin, Director, First-Year Algebra via Applications
Development Project, Department of Education, University of
Chicago, 5835 S. Kimbark Avenue, Chicago, IL   60637.

Library of Congress Cataloging in Publication Data:

First-Year Algebra via Applications Development Project.
  Algebra through applications, with probability and
statistics.

  Includes indexes.
  1. Algebra. 2. Probabilities. 3. Statistics.
I. National Council of Teachers of Mathematics.
II. Title.
QA152.2.F49 1979   512.9   78-23568
ISBN 0-87353-134-5 (set)
     0-87353-135-3 (part I)
     0-87353-136-1 (part II)

TABLE OF CONTENTS

CHAPTER 1: SOME USES OF NUMBERS

CHAPTER 2: PATTERNS AND VARIABLES

CHAPTER 3: ADDITION AND SUBTRACTION

CHAPTER 4: MULTIPLICATION

CHAPTER 5: MODELS FOR DIVISION

CHAPTER 6: SENTENCE-SOLVING

CHAPTER 7: LINEAR EXPRESSIONS AND DISTRIBUTIVITY, Part I

CHAPTER 8: LINEAR EXPRESSIONS AND DISTRIBUTIVITY, Part II

CHAPTER 1

SOME USES OF NUMBERS

Lesson 1

Some Uses of the Natural Numbers

The numbers

one, two, three, four, five, ..., one million, one million

and one, ... are called the <u>natural numbers</u>. The symbols

1, 2, 3, 4, 5, 6, 7, 8, 9, 10, 11, ...

are used in all countries and by most people on Earth.

The natural numbers have many kinds of uses.

First use of natural numbers: counting

You probably learned the natural numbers by counting 1, 2, 3,

... . If you had enough time, you might be able to count as high as

eleven million, three hundred
forty-seven thousand, four $\qquad$ 11,347,429
hundred twenty-nine.
$\qquad\quad\uparrow$ $\qquad\qquad\qquad\uparrow$
in the English language $\qquad$ in the decimal system
$\qquad\qquad\qquad\qquad$ (a mathematical language)

Here the mathematical language is much simpler than English. So

it is more often used.

Few people count from 1 to very large numbers. But large

natural numbers are still found in counting. The Constitution of

the United States requires that a census count be taken every 10 years. According to this count, there were, as of April 1, 1970,

204,765,770 people in the U.S.

At the same time, a United Nations estimate was

3,631,797,000 people in the world.

Large numbers like these are seldom counted. They are estimated. Here are some situations where estimation is necessary.

Ecology: How many whales are there?
Business: How many cars were sold last year?
Hobbies: How many stamps in a large collection?

Second use of natural numbers: identification and coding

You may have an ID number in your school. You have a phone number at home. If you work, you should have a Social Security number. Usually license plates have numbers on them. So do TV channels. These are all examples of natural numbers being used for _identification_.

Zip codes also identify. But like many other identifications, the numbers are coded. The author's zip code 60637 is coded as follows:

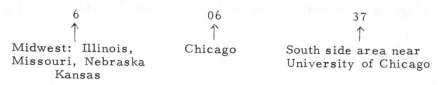

6	06	37
↑	↑	↑
Midwest: Illinois, Missouri, Nebraska Kansas	Chicago	South side area near University of Chicago

A school ID number may be coded by sex or your year in school. All credit card numbers are coded.

Third use of natural numbers: ordering

The words 1st, 2nd, 3rd, 4th, 5th, ... and so on show that the natural numbers are also used for ordering. This order is usually found by counting. A 5th-place finish means there are 4 ahead. But if there are many "2nd prizes," the 2 shows order and identifies, and is not found by counting.

Other uses of natural numbers

Natural numbers are also used in

measuring:	It is $23°$ outside.
comparison:	This book is 2 times as thick as that one.
locating:	I live at 1322 Maple.
scoring:	Did you get a 90?

These uses need more numbers than natural numbers and are described in the next two lessons.

Questions covering the reading (You should always read the entire lesson before you try to answer these questions.)

1. Name three natural numbers.

2. Which number is closest to the population of the world?

   (a) 400 million   (b) 400 billion   (c) 4 billion

3

3. Name three common types of uses of natural numbers.

4-9. Which type of use of natural numbers applies in each situation?

4. Population

5. Telephone number 555-1234

6. First floor

7. Number of bicycles sold last year

8. Zip code

9. 2nd place award

10. Name two uses of numbers which need more than natural numbers.

11. Is <u>three</u> a natural number?

12. Name a counting situation in which you would need to estimate.

13. What is a census?

14. Write in English: 2,473,169,008

15. Name one advantage of using mathematical symbols rather than words to stand for numbers.

Questions testing understanding of the reading

1. Name a number which is not a natural number.

2. Is there a largest natural number?

3. Is $\frac{1}{20}$ a natural number?

4-9. (a) Identify every natural number in the sentence. (b) Tell whether the natural number is used for counting, ordering, identification, or in some other way. (It is possible that the same number has more than one use.)

4. Tune in Channel 7 for the best movies.

5. A Boeing 747 can seat more than 300 passengers.

6. Sunday, December 7, 1941, was the day that Japan bombed Pearl Harbor, a U.S. naval base, with about 4500 American casualties.

4

7. On August 6, 1945, the U.S. dropped an atomic bomb on Hiroshima, Japan, and it is estimated that from 80,000 to over 200,000 Japanese died from either the bomb or radiation exposure.

8. Those two girls won 3rd place awards.

9. He is the only one in his family to have a 10-speed bike.

Questions for discussion and opinion

1. Why do you think the numbers 1, 2, 3, ... are called <u>natural</u> numbers?

2. Suppose you began counting out loud: one, two, three, ... How many numbers do you think you could count in one hour? How many in one year?

3. Conservationists and hunters often want to know how many animals of a particular kind are in an area. How do you think this might be done?

4. Why would anyone want to count whales?

5. The Post Office claims that "zip codes speed the mail." Do you think this is true? Back up your answer.

6. The total U.S. 1970 population 204,765,770 reported by the U.S. Bureau of the Census almost has to be inaccurate. Why?

Lesson 2

Rational Numbers and Measurement

If you divide 2 by 3, you get the number $\frac{2}{3}$. The number $\frac{2}{3}$ is not a natural number. Nor can $\frac{2}{3}$ arise as the answer to a counting problem. So there are numbers other than natural numbers. These numbers can have different uses.

If a number can be written as a fraction with natural numbers in its numerator and denominator, then it is a <u>rational number</u>. It is easy to tell that the following 4 numbers stand for rational numbers.

$$\frac{2}{3} \qquad \frac{110}{47} \qquad \frac{4,623,471}{2} \qquad \frac{83}{1}$$

Some things which do not look like fractions may still represent rational numbers.

<u>Type</u>	<u>Example</u>	<u>Written to show that the number is rational</u>
natural number	5	$\frac{5}{1}$
ratio	11:9	$\frac{11}{9}$
ending decimal	3.214	$\frac{3214}{1000}$
per cent	6%	$\frac{6}{100}$
"mixed" number	$14\frac{1}{2}$	$\frac{29}{2}$

The same rational number may be written in many different ways...

$$\frac{3}{4} \quad = \quad 3:4 \quad = \quad .75 \quad = \quad 75\% \quad = \quad 3 \div 4$$

| as fraction | as ratio | as decimal | as percent | as quotient |

... and as many different fractions.

$$\frac{3}{4} \ = \ \frac{6}{8} \ = \ \frac{9}{12} \ = \ \frac{30}{40} \ = \ \frac{75}{100} \ = \ \frac{3,000,000}{4,000,000}$$

Why are there so many ways of writing rational numbers? Because there are so many applications. Sometimes percents are easier. Sometimes decimals are easier.

First use of rational numbers: measuring

Time, distance, temperature, rates, costs--these and many other quantities are measured using rational numbers. Here are examples. The rational numbers are underlined.

1.   That girl is $3\frac{1}{2}$ years old.

2.   She makes $2.05 per hour working at that store.

3.   Normal body temperature is 37° C.

4.   As of 1975, the world's record in the women's 100-meter dash (about 110 yards) is 10.8 seconds, set in 1973 by Renate Stecher of East Germany.

There are two ways to tell if something is being measured. First, there will usually be a unit of measurement. Some common units are dollars, seconds, years, feet, meters, and so on. Second, in a measurement you can often get a little more accurate. For example, time can be done in years, months, days, hours, minutes,

7

seconds, parts of seconds, and each unit is smaller than the previous one.

Second use of rational numbers: scoring

Scoring is a special kind of measuring. Although test scores are usually natural numbers, you can get $\frac{1}{2}$ points. In diving, scores like 7.5 are used. In gymnastics, scores like 9.25 occur.

Third use of rational numbers: locating

Another special kind of measuring is locating. The address 312 Center Street locates a house. The number 9:30 (short for $9\frac{30}{60}$ hours) locates a time. Library books are often located by the Dewey Decimal System--in this system, the number 510.1 leads you to mathematics books.

Questions covering the reading

1. Give an example of a fraction.

2-5. Write each number as a fraction with natural numbers in the numerator and denominator.

2. 7:10   3. .42   4. 50 per cent   5. $2\frac{3}{4}$

6. In what way should a number be written to show that it is a rational number?

7. Which numbers of Questions 2-5 are rational numbers?

8. Why are there so many ways of writing rational numbers?

9-11. Give an example of a rational number used in

9. measurement   10. locating   11. scoring

12-17. Tell whether the number in the sentence is used for measuring, scoring, or locating. (It is possible that more than one use is given.)

12. That house is 16.3 miles down the road.

13. A bowling ball weighs about $7\frac{1}{4}$ kilograms.

14. The football team won its game by 12 points.

15. I'm hoping for a 90 on that test.

16. That recipe requires $1\frac{1}{2}$ teaspoons of salt.

17. Class begins at 10:25.

Questions testing understanding of the reading

1. Can a rational number be bigger than 1?

2. Is every natural number also a rational number?

3. Write three other fractions which equal $\frac{4}{5}$.

4. Write three other fractions which equal $\frac{11}{6}$.

5-12. Give an example of a rational number and a unit (if possible) used in measuring:

5. temperature.   6. length or distance.
7. age.   8. weight.
9. area.   10. speed.
11. cost.   12. height.

13-20. Complete each row of the table by writing the given number in all of the ways not given.

	fraction	ratio	decimal	percentage	quotient
13.	$\frac{4}{5}$				
14.		1:20			
15.					40 ÷ 8
16.				67%	
17.			.24		
18.	$\frac{1}{7}$				
19.			1		
20.			3.8		

Skill review

If you had trouble with the arithmetic in the above problems, you may need a skill review.

Directions: Try the first 5 problems in each group. Then look at the answers. (They are upside down on page 13.) If you have 4 or 5 right, go on to the next type of problem. If you get 1, 2, or 3 correct, then you need more practice. You should then try the last 5. Then look at the answers to these. If you do not have 4 or 5 right, ask your teacher for help.

Write each ratio as a fraction.

1. 3:2 2. 11:7 3. 412:610 4. 1000 to 1 5. 100000:1
6. 2:3 7. 1 to 8 8. 1.2:1 9. 1:1 10. 400:7

Write each decimal as a fraction.

11. .5 12. .23 13. 7.23 14. 8.3 15. 103.2
16. 3.14 17. 1.41 18. .9 19. 6.4 20. 109.87

Write each % as a fraction.

21. 6% 22. 1% 23. 50% 24. 250% 25. .3%
26. 20% 27. 100% 28. 1.5% 29. 28% 30. 311%

Choose the one number which is not equal to the other two.

31. $\frac{1}{3}$, $\frac{2}{6}$, $\frac{3}{10}$ 32. 1%, .1, .01   33. $\frac{5}{10}$, 50%, .05

34. 10:1, 10, 100%   35. 3.1, $3\frac{1}{10}$, 31%   36. $\frac{9}{12}$, $\frac{12}{15}$, $\frac{3}{4}$

37. 3%, $\frac{30}{100}$, $\frac{3}{100}$ 38. 6:3, 200%, $\frac{1}{2}$ 39. 6.4, 64%, .64

40. $\frac{1}{4}$, 25%, 2.5

Important skill - converting fractions to decimals and to percent.

Example 1: To convert $\frac{3}{20}$ to a decimal, recognize that $\frac{3}{20}$ is
$3 \div 20$.

Now divide:

$$\begin{array}{r} .15 \\ 20\overline{)3.00} \\ \underline{2\ 0} \\ 1\ 00 \\ \underline{1\ 00} \end{array}$$

So $\frac{3}{20}$ = .15

To convert .15 to percent, recognize that .15 is $\frac{15}{100}$. So .15
is 15%.

Example 2: Some fractions are not equal to ending decimals.

Consider $\frac{11}{7}$.

$$\begin{array}{r} 1.571 \\ 7\overline{)11.000} \\ \underline{7} \\ 4\ 0 \\ \underline{3\ 5} \\ 50 \\ \underline{49} \\ 10 \\ \underline{7} \\ 3 \end{array}$$

So $\frac{11}{7}$ = 1.571...

= 157.1...%

You could go longer.

<u>Skill review</u>.   (Answers at end of lesson.)

Approximate each fraction by a two-place decimal.

41. $\frac{7}{20}$ 42. $\frac{4}{3}$ 43. $\frac{61}{100}$ 44. $\frac{13}{40}$ 45. $\frac{81}{7}$

46. $\frac{11}{6}$ 47. $\frac{51}{80}$ 48. $\frac{3}{17}$ 49. $\frac{1}{7}$ 50. $\frac{18}{100}$

Convert to %.

51. $\frac{6}{10}$ 52. $\frac{4}{100}$ 53. $\frac{3}{5}$ 54. $\frac{1}{6}$ 55. $1\frac{3}{10}$

56. $\frac{4}{10}$ 57. $\frac{1}{100}$ 58. $\frac{3}{50}$ 59. $\frac{5}{6}$ 60. $2\frac{1}{10}$

Answers to skill review

1. 3/2   2. 11/7   3. 412/610   4. 1000/1   5. 100000/1
6. 2/3   7. 1/8   8. 1.2/1 or 12/10   9. 1/1   10. 400/7
11. 1/2   12. 23/100   13. 723/100   14. 83/10
15. 1032/10   16. 314/100   17. 141/100   18. 9/10
19. 64/10   20. 10987/10000   21. 6/100   22. 1/100
23. 50/100 or 1/2   24. 250/100 or 5/2   25. .3/100 or 3/1000
26. 20/100 or 1/5   27. 100/100 or 1/1   28. 1.5/100 or 3/200 or
15/1000   29. 28/100   30. 311/100   31. 3/10   32. .1
33. .05   34. 100%   35. 31%   36. 12/15   37. 30/100
38. 1/2   39. 6.4   40. 2.5   41. .35   42. 1.33
43. .61   44. .32   45. 11.57   46. 1.83   47. .63
48. .17   49. .14   50. .18   51. 60%   52. 4%
53. 60%   54. 16%   55. 130%   56. 40%   57. 1%
58. 6%   59. 83%   60. 210%

Lesson 3

Number Lines and Bar Graphs

Measurement and locating using rational numbers is helped and pictured by <u>number lines</u>. These lines are used in many situations.

Thermometer: Ruler:

Other than counting, no measurement can be exact. But great accuracy is often desired. Pollution is often measured in particles per million molecules of air. Metalworking requires accuracy to .001 inch and less.

Here are four number lines. The position of the number 3.4 is in the same place on each line. The other numbers indicate the <u>scale</u> of the number line.

From far away

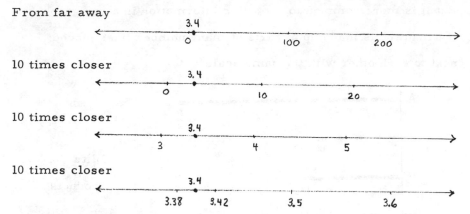

10 times closer

10 times closer

10 times closer

Decimals are usually used on number lines because they make it easy to locate the position of this number. But fractions or percents can be found on some number lines. Here is a number line with fractions above the line, percents below.

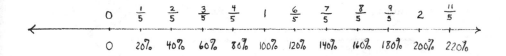

Suppose that the Blitz Co. sold 280 of product A, 98 of product B, and 123 of product C. This data could be pictured on a number line.

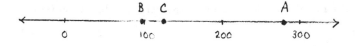

15

But it is more common to show this information in a <u>bar graph</u>.
A bar graph is made up of parts of many number lines placed
next to each other with the same scale.

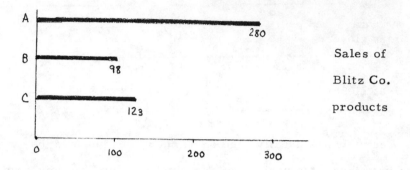

Sales of

Blitz Co.

products

On all number lines and bar graphs in this book, there is a
uniform scale. The bar graph below is crossed out because it
has been incorrectly drawn. The 98 and 123 are not to the same

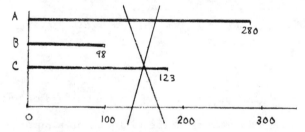

scale. When a graph does not have a uniform scale, it is difficult
to make judgments from it. So you should mark scales carefully
on a graph.

16

Questions covering the reading

1-2. Trace this number line. Then graph the given numbers on the line. Label the points graphed.

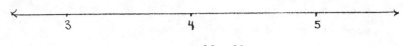

1. 3. 5, 3. 6

2. $\frac{11}{3}$, $\frac{19}{6}$

3-4. Trace this number line. Then graph the given numbers on the line. Label the points graphed.

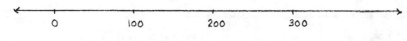

3. 130, 290

4. 20.1, 201

5-8. Graph the four given numbers on the same number line. Use a uniform scale. Label the points graphed.

5. $\frac{1}{2}$, $\frac{1}{3}$, $\frac{1}{4}$, $\frac{1}{10}$

6. 1500, 250, 750, 2000

7. $3\frac{1}{10}$, 3. 001, 3. 1, 3

8. $1\frac{2}{5}$, 1. 6, 2, 180%

9. Estimate your body temperature if the fluid in the thermometer goes until points A, B, C, D, E, or F.

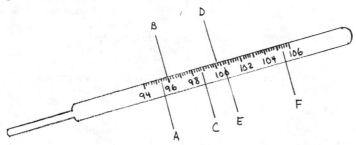

10. Why is a uniform scale helpful on a number line?

11. What is a bar graph?

17

12-13. Draw a bar graph showing the given information. Use a uniform scale.

12. Sales of 3 products in a store on a given day:

        product A: 12 items

        product B: 9 items

        product C: 25 items

13. Approximate U.S. population in four censuses:

      1800 census: 5 million   1850 census: 23 million

      1900 census: 76 million   1950 census: 151 million

Questions testing understanding of the reading

1-2. Draw a bar graph with the given information:

1. Circulation of 4 leading magazines in U.S. (data from 1971)		2. Main types of accidental deaths (data from 1971)	
Readers Digest	17, 827, 661	Motor Vehicle	54, 700
TV Guide	16, 410, 858	Falls	17, 900
Woman's Day	8, 191, 731	Burns	6, 700
Better Homes & Gardens	7, 996, 050	Drowning	7, 300
		Firearms	2, 400
		Poisons	5, 100

3-6. Graph all four numbers on the same number line.

3. Air distances from Chicago to:

    (a) Miami   1188 miles   (b) Los Angeles 1745 miles
    (c) Honolulu 4251 miles   (d) Paris   4148 miles

4. Cost of hair brushes (not including tax):

    (a) 1 brush   .49   (b) 2 brushes   .98
    (c) 3 brushes 1.47   (d) 4 brushes 1.96

5. Four students' measurements of the length of a board: (Your graph should leave a lot of space between the numbers 21 and 22.)

    (a) Wanda $21\frac{1}{2}$"   (b) Xenophon $21\frac{3}{8}$"

    (c) Yvonne   21.4"   (c) Ziggie $21\frac{7}{16}$"

6. Some U.S. Government expenditures (in dollars) in 1972: (Source: U.S. Treasury Department)

   (a) Defense   78,152,313,000   (b) Social Security   64,508,150,000
   (c) Space   3,423,939,000   (d) Health   16,969,513,000

7. In Question 5, which student was probably furthest off?

8. Put the data of Question 6 into a bar graph.

9. Put the data of Question 3 into a bar graph.

10. Many FM radios have a number line which goes from 88 to 108 and whose scale is not uniform. Of what use is this number line?

11-16. Name two numbers which are between

11. 2.5 and 2.6   12. 0 and .01   13. $\frac{1}{9}$ and $\frac{1}{10}$

14. $\frac{10}{23}$ and $\frac{11}{23}$ 15. 980.31 and 980.32

16. .33 and $\frac{1}{3}$

Lesson 4

Rational Numbers and Comparison

You can compare two numbers by dividing one of the numbers by the other. The answer is called the quotient. If you started with rational numbers, the quotient will be a rational number. The numbers you compare may be of the same kind or of different kinds.

19

<u>Comparing quantities of the same kind: ratios simplified</u>

In this type of comparison, the quotient does not have a unit. So it is not a measurement. It is a <u>ratio</u>. Ratios are often written as percentages. Here are three examples. The symbol ≈ used in Example 1 means "is approximately equal to."

<u>Sentence</u> (the number used as a comparison is underlined) | <u>Explaining the comparison</u>

1. In the 1972 Presidential Election, Nixon won about <u>61%</u> of the vote.

$$\frac{\text{number of Nixon votes}}{\text{total number of votes}} \approx .61$$

2. A class with 30 students has $1\frac{1}{2}$ times as many students as a class with 20 students.

$$\frac{30 \text{ students}}{20 \text{ students}} = 1.5$$

3. Prices are down <u>20%</u>.

$$\frac{\text{amount of decrease}}{\text{original price}} = .20$$

Suppose you want to compare several parts to a whole. Think of an entire circle as 100%. Then half the circle is 50%. Half of that is 25%. In this way, you can picture any number between 0 and 1.

20

Numbers can be combined to make a <u>circle graph</u>. Here is a summary of the 1972 Presidential election results placed on a circle graph.

	Votes	% of total
Nixon	47,167,319	60.7
McGovern	29,168,509	37.5
Others	1,345,633	1.8
Total	77,681,461	100.0

If percentages are not given, here is how you can make a circle graph. (1) Add the numbers to get a total. (2) By division, compare each number to the total. (3) Write the quotients as percentages. (4) Divide the circle into regions picturing the percentages.

<u>Comparing quantities of different kinds: rates</u>

When dividing quantities measuring different things, the quotient which results is a <u>rate</u>. The word "per" is often found in the unit.

<u>Sentence</u> (the rate is under-lined)	<u>Explaining the comparison</u>
4. We drove 100 kilometers in 2 hours, an average speed of <u>50</u> kilometers per hour.	$\dfrac{100}{2} = 50$
5. In that school, there are about <u>23.7</u> students per class.	$\dfrac{\text{no. of students}}{\text{no. of classes}} \approx 23.7$

21

6. Ty Cobb had a lifetime
   batting average of nearly
   .367, the Major League
   record.

   $\dfrac{\text{number of hits}}{\text{number of at-bats}}$ ≈ .367

To summarize, these are the major **uses** of numbers.

1. counting
2. ordering
3. identification
4. measurement (including scoring and locating)
5. comparison (including ratios and rates)

The same number may combine two or more of these uses.

Questions covering the reading

1. The symbol " ≈ " means _____.

2. <u>Multiple choice</u>. In 1972, McGovern's part of the vote ≈
   (a) 20%   (b) 30%   (c) 40%   (d) 50%   (e) 60%

3. In a school election, 54% of the votes went to the winner.
   What two quantities were being compared to arrive at the
   "54%"?

4. Comparing what two quantities leads to a "batting average"?

5-10. Estimate the percentage of the circle which is shaded.

5.   6.   7.   8.   9.   10.

22

11-12. Estimate the percentage for each region in these circle graphs.

11.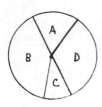

12.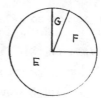

13-14. Picture all of the given percentages on a single circle graph.

13. Political party affiliation in a certain county (not actual data)

Democrat	36%
Republican	29%
Other	4%
Independent	31%

14. Women over the age of 14 in the U.S. (Based on 1970 Census)

Single, never married	22%
Married	61%
Divorced	4%
Widowed	12%

15. In Exercise 14, why do the percentages not add to 100%?

16. What are the four steps in constructing a circle graph of some numbers?

17-22. Make up a situation which could lead to the given rate. (Hint: Example 4 on page 21 shows how you could get 50 kilometers per hour.)

17. 40 words per minute

18. 1 minute a question

19. 27 students per class

20. 20 kilograms per suitcase

21. 76¢ a can

22. $2\frac{1}{2}$ children per family

23. Name the five major uses of numbers.

Questions testing understanding of the reading

1-4. Divide one of the given numbers by the other. What does the fraction stand for?

1. Six cans of orange juice cost $1.50 recently.

2. A pad of 100 pieces of paper is about 1 cm thick.

3. There are about 8 kilometers in 5 miles.

4. 27 of the 50 states are east of the Mississippi.

5-8. Each sentence contains a rational number found by comparing. Name the number and the two quantities being compared.

5. That dress uses twice as much material as this one.

6. This table is only three-fourths as long as that one.

7. Workers in plant X would like a 10% raise this year.

8. Prices slashed 30%!

9-10. Picture the given information on a single circle graph. Follow the 4 steps given on page 21.

9. A typical day?

School	7 hours
Home	12 hours
Outside	3 hours
Somewhere else	2 hours

10. A typical High School population?

300 Freshmen
275 Sophomores
225 Juniors
200 Seniors

11. Make up your own example like those in Questions 17-22 on page 23.

Throughout this book, the C next to a problem means that calculators will be helpful in doing the problem.

12-13. Make one circle graph which shows all of the given numbers.

12. $\frac{1}{2}, \frac{1}{3}, \frac{1}{6}$

C 13. $\frac{1}{7}, \frac{2}{7}, \frac{4}{7}$

C 14-15. Make a circle graph using the given data.

14. Land Area of Continents (sq. mi.)

Asia	16,988,000
Africa	11,506,000
N. America	9,390,000
S. America	6,795,000
Europe	3,745,000
Australia	2,968,000
Antarctica	5,500,000

15. Population of Continents (1970)

2,048,898,000
344,000,000
314,970,000
186,013,000
637,943,000
12,200,000
0

1. For a circle graph, would you prefer to be given data in decimals, fractions or percents?

2. The data in Question 14 above are given in square miles. Would the circle graph look any different if the measurements were in square kilometers?

Summary Questions

1-5. Identify all of the numbers in the given sentences. Tell whether the number is a natural number or not. Tell whether the number is being used for counting, ordering, identification, measuring, or comparison.

   Example: The 4 items together cost $59.26.

   Answer: 4 - a natural number - used for counting
         59.26 - not a natural number - used for measuring

1. That car gets 15.6 miles per gallon.

2. About $\frac{1}{86}$ of all births are twins.

3. The Carson family has lived at 123 Elm for the past 12 years.

4. (from the <u>Guiness Book of World Records</u>) "The largest diamond ever discovered was a 3,106 metric carat (over 1 1/4 lb.) stone by Captain M.F. Wells, in the Premier Mine, Pretoria, South Africa, on January 26th, 1905."

5. She scored a 37.22 on that dive.

6-7. Follow the directions of Questions 1-5. These are quotes from President Ford in 1974.

6. "...inflation is our Public Enemy No. 1."

7. "We're going to hold the line on spending and meet the target of a budget under $300 billion in fiscal 1975."

8-11. These paragraphs are taken from the <u>Chicago Sun-Times</u> of August 25, 1974. Follow the directions of Questions 1-5.

8. "A window-washer survived a four-story fall while his partner dangled from a safety line Saturday at East Point Condominiums, 6101 N. Sheridan... 'The two were washing windows between the sixth and seventh floors of the 42-story building when the cable holding their scaffold slipped,' said 27th Fire Battalion Chief Robert Haig."

9. "Sale! Back-To-School Supplies. Ring Binder Set. Reg. 2.69. Binder with booster metal, 3-ring. Contains 22 sheets, dictionary, zipper carry-all. 2.29."

10. "For the ninth time in 11 starts this year, David Pearson has won the front row pole position for a major stock car race. The 39-year-old $1 million career winner from Spartanburg, S.C. wheeled a Mercury around the two-mile oval Michigan International Speedway to claim the honor of leading a 36-car starting field to the post in Sunday's $82,200 Yankee 400."

11. "The long dress...is the Ibo style and takes a minute or two to wrap. You need eight yards of 48-inch wide fabric. Begin at the waist, making sure you cover the ankles, and keep winding until you reach the shoulder, when you tuck the excess fabric in the back. Spread your feet apart 24 inches before you begin to wind so you'll be able to walk easily."

<u>Follow these directions</u>

A. Find a page from a daily newspaper. Write down the name of
    the newspaper, the date, and the page.

B. Look for numbers on the page. Choose 10 numbers which have
    some variety. You may want to circle the numbers.

C. Copy and fill in the following table. If you picked page 14, a
    possible first entry is done for you.

	Number	Is the number a natural number?	How is the number used?
1.	__14__ (page number)	Yes	Counting, Identification
2.			
3.			
4.			
5.			
6.			
7.			
8.			
9.			
10.			

D. How many numbers seem to be on a typical page of a newspaper?
    _____ What are the most common uses of these numbers?

Lesson 5

TV Ratings and Sampling

It costs money to make a TV program. Most TV shows are
paid for by large companies like IBM, Revlon, Chrysler, and
Kellogg. The companies do this in order to sell their products.
So they want lots of viewers.

How does a company know if a lot of people are watching its
program? It receives its program's ratings from other companies
whose business it is to estimate how many people are watching
each program. You know that programs are dropped because of
low ratings. So ratings are important to entertainers as well as
companies.

TV ratings are determined by a process called sampling.
Here are the steps in that process.

Step 1. The set of all things that could be studied is described.
This set is called the population.

For TV ratings, the population
is the set of all people who watch
TV. It is not easy to describe
this population. But the census
helps. At right is a circle graph
showing a distribution of ages in
the U.S. population in 1970.

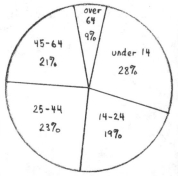

Step 2. A subset of the population is very carefully selected.

This subset is called a sample. It usually contains

people in from 1000 to 3000 families. The sample is

designed to be very much like the larger population.

Step 3. The sample is polled to find out what TV programs are

watched. The polling is sometimes done by phone call.

Sometimes families are asked to write down what programs

they watched. Sometimes an electric monitor is placed on

each set.

Step 4. Analyze results for the

sample. In a TV survey,

one might get data as

listed at right.

At 8:45 PM on September 12th, in the sample:
24. 3% were watching Program A
15. 7% were watching Program B
10. 1% were watching Program C
11. 3% were watching others
38. 6% were not watching TV

The sample percentages are known as the TV ratings.

Program A would have a rating of 24. 3, Program B 15. 7,

and so on.

Step 5. Percentages for the population are estimated from the per-

centages from the sample. The estimate might be that 22%

to 26% of the population was watching Program A. Of

course, the estimate might be wrong. But such estimates

are correct at least 95% of the time.

29

There are always at least five steps in sampling procedures.

Population	Sample

1.   Identify the population. $\longrightarrow$ 2.   Carefully select a sample
$\qquad\qquad\qquad\qquad\qquad\qquad\quad\downarrow\quad$ like the population.

$\qquad\qquad\qquad\qquad\qquad\qquad\quad$ 3.   Study sample.
$\qquad\qquad\qquad\qquad\qquad\qquad\quad\downarrow$
5.   Apply results from $\longleftarrow$ 4.   Get results from sample.
     sample to gain
     information about the
     population.

Sampling is used in virtually every walk of life.

1.   Consumer testing.   You want to know whether A & P's "mixed nuts" has an accurate picture on its label.   (Perhaps the picture is more attractive than the nuts inside.) So you purchase some jars in order to count the nuts.

    Population: The set of all jars of A & P mixed nuts manufactured.

    Sample: The set of all jars purchased by you.

In this example, the population did not consist of people.

2.   Medicine.   Doctors want to test a drug to fight lung cancer. Suppose everyone with lung cancer was given the drug.   Then no one would know how good the drug was!   Two samples would be chosen.

Population: The set of all people with lung cancer

Sample 1:   People with lung cancer who are given the drug

Sample 2:   People with lung cancer who are not given the
drug but are studied

3. Election prediction. Candidates and newspapers always try to predict how an election is going. So they poll.

Population: The set of all registered voters

Sample: The set of all people polled

4. Gambling. You wonder whether a dime is weighted. So you toss the coin 100 times.

Population: The set of all possible tosses of the coin

Sample: Your 100 tosses

In this one the population is an infinite set. You must sample.

Sampling is used because working with the entire population would:

be too expensive (as with TV ratings).
be too hard (as in election prediction).
ruin the population (as in consumer testing).
be impossible (as in gambling).

Questions covering the reading

1. In the sampling process, what is the population?

2. In sampling, what is the sample?

3. How are TV ratings found?

4-5. In the example of TV ratings given in this lesson, what was the rating of

4. Program B   5. Program C

6. What does a TV rating of 20.1 for a program indicate?

7. What does a TV rating of 6 indicate?

8. What are the steps in using the sampling process?

9. Name four reasons why sampling is used.

10-13. Indicate why sampling would be necessary in each situation.

10. predicting results of a school election

11. telling whether a penny is balanced

12. testing the lifetime of light bulbs

13. testing a new vaccine

Questions testing understanding of the reading

1-4. Give a possible population and sample for each situation.

1. You want to know what percentage of high school freshmen can play the guitar.

2. You wish to determine whether a particular pair of dice is loaded.

3. You want to know which song is most often played on radio in the country in the week October 1-7.

4. A manufacturer wants to test a new hair spray.

C 5. Suppose the TV ratings show that at a particular time,

       program A has rating 16.4
       program B has rating 21.2
       program C has rating 12.5
       program D has rating   3.1

and 1600 people were in the sample. How many people in the sample:

(a) watched program A   (b) watched B   (c) watched C

(d) watched D   (e) were not watching television at the time

C 6. Repeat Question 5 for the ratings given in this lesson, p. 29.

7. Nancy's rating service says that its samples are usually accurate to within 3%. In an attempt to predict an election, in the sample

       48% prefer Candidate A
       52% prefer Candidate B

Should Nancy predict that Candidate B will win? Why or why not?

8. 250 phone calls are made in a particular city to help determine TV ratings. Here are the results:

       52 people watching Channel 4
       36 people watching Channel 5
       16 people watching Channel 7
       51 people watching Channel 10
       35 home but not watching
        4 have no TV
        6 home but refuse to answer the question
       50 no answer to phone call

What ratings would you give the four channels? (There is more than one way to interpret this data.)

Questions for discussion and opinion

1-4. Argue whether or not the sample is representative of the entire population.

1. In a large city, to find out which mayoral candidate might win, a pollster decides to turn to a page of the phone book and calls every person on that page.

2. A teacher stops one of his or her students in the halls to find out whether the teacher remembered to give the assignment for the next day.

3. A potential vaccine for a disease is tested on prisoners.

4. To find out how many people cheat on income tax returns, the returns of all those who earned over $50,000 are checked.

5. Why might a manufacturer of skin cream want TV ratings only for the age group 14-24?

6. Do you know anyone who has been part of a TV ratings sample?

Lesson 6

Relative Frequencies: Another Use of Ratios

An **experiment** is any situation that we wish to study. Sampling is often a part of an experiment. Tossing a coin is a type of experiment. The possible results of an experiment are called outcomes. In coin tossing, there are two outcomes: heads (H) and tails (T).

Suppose you toss a coin 150 times and get 80 heads.

$$\frac{\text{number of heads}}{\text{number of tosses}} = \frac{80}{150}$$

The ratio $\dfrac{\text{number of heads}}{\text{number of tosses}}$ is the <u>relative frequency</u> of getting

heads in tossing a coin. In general:

Definition:

The <u>relative frequency</u> of an outcome to

an experiment is the ratio

$$\dfrac{\text{frequency of the outcome}}{\text{number of repetitions of the experiment}}$$

As this lesson was being written, the author decided to

calculate relative frequencies of heads for a nickel he had in his

pocket. He tossed the nickel 80 times. (This is a sample of

all possible tosses.) Here were his results.

Tosses					Number of heads	Number of tails
1-20	HHTHH	HHHTT	THHHT	TTHHH	13	7
21-40	HHHHH	THHHT	HHHTH	THHTH	15	5
41-60	HTHHT	HTTHH	TTTTH	THTHH	10	10
61-80	TTTHH	TTTHT	THTHH	THTTH	8	12
				Totals	46	34

After 80 tosses, the relative frequency of heads is

$$\dfrac{\text{frequency of heads}}{\text{number of tosses}} = \dfrac{46}{80} = .575$$

If the coin is balanced, you would expect the ratio to be near

.50 or $\dfrac{1}{2}$. If this did not occur, it might be that (1) the coin is

weighted, (2) the author is not tossing the coin fairly, or (3) the coin has not been tossed long enough. (The number $\frac{1}{2}$ is called the <u>probability</u> of getting heads in <u>one</u> toss of a fair coin. You will study probability in Chapter 5.)

Example: Suppose a tire company tests 50 tires to see how long
they last under typical road conditions. Here are the
results they might get.

<u>Mileage until worn</u>	<u>Number of tires</u>
10,000 - 14,999	1
15,000 - 19,999	3
20,000 - 24,999	6
25,000 - 29,999	15
30,000 - 34,999	14
35,000 - 39,999	7
40,000 - 50,000	4

1. What is the relative frequency of a tire lasting over 25,000
miles?

Answer: relative frequency = $\dfrac{\text{no. of tires lasting over 25,000 miles}}{\text{total no. of tires tested}}$

$$= \frac{15 + 14 + 7 + 4}{50} = \frac{40}{50} = .8$$

2. What is the relative frequency of a tire lasting over 10,000
miles?

Answer: relative frequency = $\dfrac{\text{no. of tires lasting over 10,000 miles}}{\text{total no. of tires tested}}$

$$= \frac{50}{50} = 1 \quad \underline{\text{This event always occurred.}}$$

3. What is the relative frequency of a tire lasting over 75,000
    miles?

    Answer: relative frequency = $\dfrac{\text{no. of tires lasting over 75,000 miles}}{\text{total no. of tires tested}}$

    $\qquad\qquad\qquad\qquad\qquad = \dfrac{0}{50} = 0$ <u>This event never occurred</u>.

In summary,

(1) Relative frequencies are always between 0 and 1, inclusive.

(2) A relative frequency of 0 means that an event has never occurred.

(3) A relative frequency of 1 means that an event has always occurred.

(4) The more often an event occurs relative to the number of repetitions of the experiment, the closer its relative frequency is to 1.

<u>Questions covering the reading</u>

1. What is an experiment?

2. What are the outcomes of an experiment?

3. What is the relative frequency of an outcome to an experiment?

4-11. Suppose you toss a coin 4 times. Give the relative frequency of <u>heads</u> if exactly:

4. no tosses are heads	5. 1 toss is heads
6. 2 tosses are heads	7. 3 tosses are heads
8. 4 tosses are heads	9. no tosses are tails
10. 3 tosses are tails	11. 2 tosses are tails

37

12-19. Refer to the example on page 36. What is the relative frequency of a tire lasting:

12. between 20,000 and 24,999 miles.

13. between 10,000 and 19,999 miles.

14. under 25,000 miles.   15. 25,000 miles or more.

16. under 40,000 miles.   17. 40,000 miles or more.

18. over 70,000 miles.   19. under 70,000 miles.

20. Match each event with the possible relative frequency.

Event	Relative Frequency
(a) never occurred	(e) 1
(b) always occurred	(f) 2
(c) occurred as often as it didn't	(g) 0
	(h) $\frac{1}{2}$
(d) occurred 2 in 3 times	(i) .666...

21-24. In the author's tosses of his nickel, what was the relative frequency of tails:

21. in the first 20 tosses.   22. in the first 50 tosses.

23. in the first 2 tosses.   24. in the 80 tosses.

Questions testing understanding of the reading

1-6. Trace or draw this number line. Based on your experience, graph the relative frequency of:

1. the sun rising tomorrow.

2. the moon falling to the Earth before this day is over.

3. it raining a week from today.

4. a boy being born in the birth of a single baby.

5. twins being born in your family.

6. triplets being born in your family.

7. According to the <u>Columbia Viking Desk Encyclopedia</u>, 1 in about 86 deliveries in the U.S. results in twins. What is the relative frequency of:

    (a) having twins?   (b) not having twins?

8. You toss a coin 50 times. 30 heads occur. Give the relative frequency of:

    (a) heads.   (b) tails.

C 9. In 1968, there were 1,796,326 male births and 1,705,238 female births in the U.S. According to this data, what is the relative frequency of female births? Does this answer agree with your answer to Question 4?

C 10. In 1971, approximately 1,920,000 people died in the United States. A sample was made of 10% of these deaths. In the sample, heart disease caused 74,017 deaths. (a) What is the relative frequency of a person dying from heart disease? (b) From this, about how many people in the U.S. died of heart disease in 1971?

C 11. (Refer to Question 10.) In the sample, 11,099 people died of accidents. What is the relative frequency of a person dying from an accident? (About half the accidents were auto accidents.)

12-13. A <u>die</u> (plural "dice") is a cube with numerals or spots on it; the six sides indicate the six possible outcomes 1, 2, 3, 4, 5, 6. If a die is tossed and lands so that 4 is on top, then a "4" has been thrown. This is the outcome pictured at right. If <u>two</u> dice are tossed, the outcome is found by adding the numbers on top, so 2, 3, 4, 5, 6, 7, 8, 9, 10, 11 and 12 are possible.

12. Toss one die 60 times and record the outcomes. Calculate the relative frequency of each possible outcome. Do you think your die is balanced?

13. Toss two dice 100 times and record the outcomes. Calculate all relative frequencies. Why, under normal circumstances, will 7 come up more often than 2?

14. Five girls are entered in a
gymnastics meet. Who goes first
is important, so a spinner is
built like that at right. Do you
think this is a fair way to decide?
Why or why not?

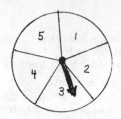

Questions for discussion and exploration

1. Refer to the example of page 36. If you were the tire company, for how long would you guarantee these tires?

2. Refer to Question 8 p. 39. Do you think the coin could possibly be balanced? Toss a thumb tack 100 times. What is your relative frequency that the tack lands on its side? If possible, compare your results with others.

3. Build a spinner like that in Question 14.

   (a) Hold your spinner level to the ground. Spin it 50 times to test its fairness. Calculate the relative frequencies. Do you think your spinner is fair?

   (b) Hold the spinner vertically with "1" at the top. Compare your results with those of part (a). Is this a fair way to hold your spinner?

4. Try to find more recent data than that given in Questions 9-11. (An almanac is a possible place to look.)

Lesson 7

Negative Numbers: Numbers to Indicate Direction

A business may make money. Or it can lose money. Or it
can break even. Here are some of the possibilities.

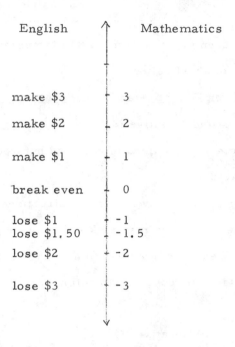

English	Mathematics
make $3	3
make $2	2
make $1	1
break even	0
lose $1	-1
lose $1.50	-1.5
lose $2	-2
lose $3	-3

The numbers at the right of the line help to describe the
situation without using words. Higher numbers on the graph
mean better business. The − sign stands for opposite of. The
opposite of making $3.25 is losing $3.25. So the opposite of
3.25 is -3.25.

The numbers with ⁻ signs are called <u>negative numbers</u>. Many English phrases are used with negative numbers.

⁻3 = negative 3	BEST WAY
⁻3 = the opposite of 3	OK - mathematically correct but a little long
⁻3 = minus 3	very common way but confusing because there is no subtraction

You have seen negative numbers in temperatures. They are found in many other situations.

In bowling on TV, <u>⁻12</u> stands for "<u>behind by 12 pins</u>." The symbol "<u>+12</u>" called "positive 12," stands for <u>ahead by 12 pins</u> (+12 is the same as 12). Zero is not used but would mean that the bowlers are even.

When a situation has two opposite directions, negative numbers may be appropriate. You may pick either one of the directions to be positive. Then the other direction is negative. Zero stands for the neutral state. By custom, directions are usually chosen in the following way.

Situation	Negative direction	Zero	Positive direction
Elevation on Earth	below sea level	sea level	above sea level
business	loss	break even	profit
games	behind	even	ahead
time	before	now	after
trade	giving out	no trade	taking in
savings account	withdrawal	no transaction	deposit

42

On a horizontal number line, negative numbers are almost always placed at the left. Numbers at the left are smaller than numbers to the right.

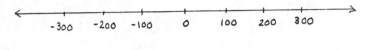

smaller numbers   larger numbers

The symbol < means "is smaller than" or "is less than." The symbol > means "is greater than" or "is larger than."

2 < 3	2 is less than 3
-100 < -80	-100 is less than -80
4 > 0	4 is greater than 0
$5 > -10\frac{1}{2}$	5 is greater than $-10\frac{1}{2}$

The symbol always points to the smaller number and opens up to the larger. If the numbers stood for profit or loss, the larger numbers (even if losses) mean better business.

Inequality and equality symbols can be combined. The symbol ≤ means "is less than or equal to."

$$5 \le 5 \qquad -47 \le -10 \qquad 4 \le 312$$

The symbol ≥ means "is greater than or equal to."

$$3 + 3 \ge 6 \qquad 126 \ge 125.3 \qquad 90 \ge -91$$

It is also possible to use more than one sign in a sentence. For example: 2 < 3 < 5 means "3 is between 2 and 5." You could also write 5 > 3 > 2. However, the < and > signs are never combined in a sentence.

1-4. Translate into English.

1. -3 2. $-8 < 4$ 3. $-8\frac{1}{2}$ 4. $6 > -1$

5-8. Translate into mathematics.

5.   negative 7   6. four is less than 10

7.   the opposite of 8   8. three is greater than negative three

9-10. Draw a horizontal number line. Graph the numbers on that line.

9. $-150,\ 230,\ -60,\ -40,\ 20$ 10. $-\frac{1}{2},\ 1\frac{1}{2},\ -3\frac{2}{3},\ -4,\ 9$

11-12. Draw a vertical number line. Graph the numbers on that line.

11.   profit of $3.20, loss of $10, loss of $6.50, break even, loss of $.25

12.   temperatures: $0°,\ -7°,\ -14°,\ -21°,\ -28°$

13.   When may negative numbers be appropriate in a situation?

14.   What is the difference between the numbers 5 and +5?

15-20. Rank the numbers from smallest to largest.

15.   0, 7, 15   16. $-3,\ 1,\ 0$

17.   100, 50, -300 18.   3.1, 3.2, 3.19

19. $-1,\ -8,\ -7$ 20. $-40,\ -90,\ 0$

21-26. True or false?

21. $13 > 6$ 22. $-1 > -2$

23. $-40.3 > -40.2$ 24. $17 < 10.346$

25. $-11 < -1$ 26. $-700 < 14$

27.   Zero separates the _____ numbers from the _____ numbers.

28-33. True or false?

28. $5 \leqslant 5$

29. $40 \leqslant -40$

30. $-6 \geqslant 10$

31. $.003 \geqslant 0$

32. $2 < \dfrac{5}{2} < 3$

33. $\dfrac{1}{4} \geqslant \dfrac{1}{3} \geqslant \dfrac{1}{2}$

Questions testing understanding of the reading

1-5. Give numbers which could be used in each situation.

1. Bowling:
   - (a) down by 20 pins
   - (b) up by 34 pins
   - (c) match is even

2. Business:
   - (a) a profit of $1,283.47
   - (b) a loss of $623.93
   - (c) a loss of $2.46

3. Elevation:
   - (a) Mt. Everest, Nepal, 29,028 feet above sea level
   - (b) level of Mediterranean Sea
   - (c) shore of Dead Sea, Israel, 1286 feet below sea level

4. Savings account:
   - (a) withdrawal of $25
   - (b) deposit of $60.00

5. U.S. Foreign Trade:
   - (a) a trade excess (more out than in) of $2 million
   - (b) a trade deficit (more imports than exports) of $5 million

6. In the televised countdown before a rocket blastoff, $\begin{smallmatrix} -1 & 23 \\ \text{min} & \text{sec} \end{smallmatrix}$ $\begin{smallmatrix} -1 & 22 \\ \text{min} & \text{sec} \end{smallmatrix}$ $\begin{smallmatrix} -1 & 21 \\ \text{min} & \text{sec} \end{smallmatrix}$ ... could be seen. (a) What did these numbers stand for? (b) What would 0 stand for?

7. In golf, par is the score a very good golfer is expected to get. Scores below par are very fine and are indicated by negative numbers. Professionals do not like to score above par. Above par scores are indicated by positive numbers. (a) If par for a course is 72, how do you think a score of 71 is indicated? (b) What would indicate a score of 78?

45

8. If a temperature of -7° falls 1°, what is the new temperature?

9-12. Fill in the blank with a number which makes the statement true.

9. $-4 <$ _____ < -3 10. $-1000 >$ _____

11. $1 >$ _____ $\geqslant -1$ 12. $0 >$ _____

13. Suppose time is measured in days and 0 stands for today. What number would stand for each day?

    (a) yesterday   (b) day after tomorrow   (c) a week ago

14. Suppose time is measured in years and 0 stands for this year. What number would stand for each year?

    (a) next year   (b) 2 years ago   (c) the year 1950

15-17. Translate each sentence into mathematics.

15. In business, a loss of $3 is better than a loss of $103.

16. In foreign trade, a trade deficit of $2 million is not as good as a trade excess of $.5 million.

17. In football, a gain of 3 yards is better than a loss of 4 yards.

18-21. Notice that $-(-3)$ is the opposite of -3, so $-(-3) = 3$. Use this idea to simplify each expression.

18. $-(-200)$ 19. $-(-(-87.2))$

20. $-(-(-(-\frac{1}{7})))$ 21. $-(-(-(-(-(-(-(1)))))))$

Questions for discussion and opinion

1. Examine the golf question #7, p. 45. Why do you think scores are related to par (-1 means 1 below par) rather than just counted as they are?

2. Make up a situation where positive and negative numbers might be used. Do not use any situation mentioned in this lesson.

Lesson 8

The Decimal System

The <u>decimal or base 10 system</u> is the system of writing numbers based on 10. For example, 432 and 2.78 are <u>decimals</u>. Notice how they are based on 10. (The raised dot indicates multiplication.)

$$432 \ = \ 4 \cdot 100 + 3 \cdot 10 + 2$$
$$2.78 \ = \ 2 + 7 \cdot \frac{1}{10} + 8 \cdot \frac{1}{100}$$

The decimal system was developed by the Hindus in the years 600-1200. Europeans found out about this system from the Arabs, who traded with the Hindus. (Arabs and Moslems had conquered and settled in parts of Europe from the 600's to the 1400's.) This is why the numerals 0, 1, 2, ... are called <u>Arabic numerals</u>.

SOME NUMERATION SYSTEMS

Arabic (decimal)	Roman	Greek	binary
1	I	α′	1
2	II	β′	10
3	III	γ′	11
4	IIII or IV	δ′	100
5	V	ε′	101
6	VI	ϛ′	110
7	VII	ζ′	111
8	VIII	η′	1000
9	VIIII or IX	θ′	1001
10	X	ι′	1010
20	XX	κ′	10100
50	L	ν′	110010
100	C	ρ′	1100100
111	CXI	ρια′	1101111
500	D	φ′	111110100
1000	M	͵α	1111101000
1500	MD	͵αφ′	10111011100
1976	MCMLXXVI	͵αϡοϛ′	11110111000
10000	X̄	͵ι	10011100010000
100000	C̄	͵ρ	110000011010100000

47

The decimal system has not been the only system ever used. The numerals I, V, X, ... were used by the ancient Romans. The Greeks used the letters of their alphabet to represent numbers. And there are other systems used today, like the binary system.

By the 1500's decimals were commonly used for natural numbers. But the stamp pictured below shows that even in 1944 not all countries used decimals. But today <u>all</u> countries use the decimal system. It is necessary in trade and communication.

1944 Inner Mongolia stamp

4 fen

It has taken about 1500 years for the decimal system to be adopted by everybody. But the system will probably be around for a long time in its present form (though maybe with different symbols). This is because, in the decimal system

(1) Only 10 symbols and a dot . are needed.

(2) The numbers most commonly used are easy to represent with just a few symbols.

(3) Addition is very easy.

(4) Multiplication is not as easy but not very hard.

(5) It is easy to tell which of two given numbers is larger or smaller.

48

In applications, decimals are used more than fractions. The small pocket calculators and the increased use of the metric system mean even more uses for decimals. (Today's calculators show all answers in decimals.)

In 1585, Simon Stevin first popularized the use of decimals for fractions. Here is how this is done. Consider $\frac{14}{11}$.

$$
\begin{array}{r}
1.2727 \\
11)\overline{14.00000}\ldots \\
\underline{11} \\
3\ 0 \\
\underline{2\ 2} \\
80 \\
\underline{77} \\
30 \\
\underline{22} \\
80
\end{array}
$$

The pattern is clear. The 2's and 7's will continue to appear. This shows that the decimal for $\frac{14}{11}$ is unending. It is an <u>infinite decimal</u>.

$$\frac{14}{11} = 1.272727272727272727272727272727272727272727\ldots$$

For practical purposes, an abbreviation is needed. To show that the "27" repeats, on and on without end, a bar — is put over it.

$$\frac{14}{11} = 1.\overline{27}$$

Here are decimal equivalents for some simple fractions. Some repeat, some do not.

$$\frac{1}{2} = .5 \qquad\qquad \frac{1}{3} = .\overline{3}$$

$$\frac{1}{4} = .25 \qquad\qquad \frac{1}{6} = .1\overline{6}$$

$$\frac{1}{5} = .2 \qquad\qquad \frac{1}{7} = .\overline{142857}$$

$$\frac{1}{8} = .125 \qquad\qquad \frac{1}{9} = .\overline{1}$$

These fractions illustrate an important property of rational numbers. <u>Decimals for rational numbers either end or repeat</u>.

Questions covering the reading

1. Show how 226 is based on 10.

2. Which number is not a decimal?
    (a) 7860 (b) 463.4 (c) $\frac{2}{5}$

3. Who developed the decimal system?

4. Who transmitted the decimal system to Europe?

5. Who was the first to use decimals for fractions? When was this done?

6. Name the 10 symbols used in the decimal system.

7-12. Give a fraction for each decimal.

7. $.33\overline{3}$ 8. $.\overline{1}$ 9. $.25$
10. $.2$ 11. $.\overline{142857}$ 12. $.125$

13. Why will decimals be used even more in the future than in the past?

14. True or False: $61.\overline{34} = 61.3\overline{43}4$

15. True or False: $61.\overline{34} = 61.3\overline{43}$

16. Pick the easiest problem and do it.
    (a) Which is larger, MCDXLVII or MCCCLXXIII?
    (b) Which is larger, 346.284 or 345.791?
    (c) Which is larger, $\frac{31}{7}$ or $\frac{22}{5}$?

17. What feature of decimals is shown by Question 16?

18. Look at Question 16. Which pair of numbers is easiest to add?

19. Name two reasons why the decimal system will probably be around for a long time.

20. (a) What division can be done to find the decimal for $\frac{2}{7}$?
    (b) Find the first 7 places of this decimal.

21. Decimals for rational numbers either _____ or _____.

Questions testing understanding of the reading

C 1-8. Find the ending or repeating decimal for each rational number.

1. $\frac{40}{7}$ 2. $-\frac{1}{3}$ 3. $\frac{7}{125}$ 4. $\frac{4}{99}$

5. $-\frac{39}{26}$ 6. $\frac{218}{5}$ 7. $\frac{1}{11}$ 8. $\frac{1}{16}$

9-12. Place a < , =, or > sign in the blank. Be careful.

9. .33 _____ .3   10. $-.\overline{88}$ _____ $-.\overline{888}$

11. 4.356 _____ $4.3\overline{5}$ 12. $71.9\overline{01}$ _____ $71.8\overline{89}$

13. If a team wins 2 of 3 games, its winning percentage is usually given as .667. What would be more accurate?

14. Suppose a batter has 4 hits in 14 at-bats. A newspaper will list the "batting average" as .286. What is it really?

15-16. Fill in the blank to make a true statement.

15. $.1 <$ _____ $< .\overline{1}$ 16. $.88 <$ _____ $< \frac{8}{9}$

Lesson 9

The Metric System

Units of measurement developed independently of the decimal system. At first they were very rough. Most units were based on the human body. (See the drawing below.) Everyone's foot had a different length! This did not pose as many problems as it might seem. All shoes were handmade. Mass production was unknown.

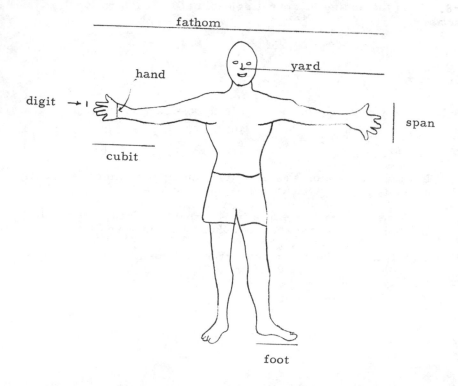

Drawing copied from <u>The Realm of Measure</u>, by Isaac Asimov

Scientific experimentation began in the time of Galileo, around 1600. In order to understand someone else's work, you had to know what the units meant. So units became standardized. One standardized system was called the English system. Inches, feet, yards, and miles measured distance. Land was measured in rods. The height of horses was (and still is) measured in hands. Depth of water was measured in fathoms. The dimensions of Noah's ark (Bible: Genesis 6:15) are given in cubits. And there are mils, leagues, digits, and spans, all measures of length.

All of these terms make things confusing. This mess was recognized by the writers of the Constitution of the United States. They wrote:

> The Congress shall have power
> ...to fix the standard of weights
> and measure.

So, in 1790, Thomas Jefferson (then Secretary of State) proposed to Congress a measuring system based on 10. (In 1786 Congress had approved our monetary system, which is based on decimals.)

At about the same time (1795), after much scientific work, a new measuring system was established in France. Like Jefferson's idea, this system was based on the decimal system. It is called the metric system because the fundamental unit of length is the meter. But the United States did not adopt Jefferson's proposal or the metric system. This was because we were historically

and emotionally tied to England. We were not as close to France.
So in the 1800's we gradually adopted the English system.

Because the metric system is so easy, country after country
adopted it. In the early 1970's even England went metric. The
United States finally officially adopted the metric system in 1975,
the last industrialized country to do so. The plan is to convert
all official signs and measurements by 1985.

In the metric system, certain prefixes have fixed meanings.
The most common are:

$$\text{milli-} = \frac{1}{1000} \qquad \text{centi-} = \frac{1}{100} \qquad \text{kilo-} = 1000$$

Thus one <u>millimeter</u> = $\frac{1}{1000}$ meter. One <u>centimeter</u> = $\frac{1}{100}$ meter.
A <u>kilometer</u> = 1000 meters. Notice that this system is based on
10, just like the decimal system. That is what makes it so easy.
These units are abbreviated m, mm, cm, and km. No periods
are used.

The basic unit of mass--also used for weight--is the <u>gram</u>.
So 1 milligram= $\frac{1}{1000}$ gram. One <u>kilogram</u> = 1000 grams. These
units are abbreviated g, mg, and kg.

In some fields, the United States has used the metric system
for a long time. Most medicine in the U.S. is done in metric.
(You can see milligrams (mg) on boxes of vitamins. You will not
see ounces.) So is most photography. Almost all science is done
in metric.

In 1959, the English units were redefined using the better metric units. Since then,

1 inch = 2.54 centimeters exactly.

There are 3 approximate conversions which you should know.

1 kilogram ≈ 2.2 pounds
1 meter ≈ 39.37 inches
1.6 kilometers ≈ 1 mile

Questions covering the reading

1. Using a ruler, draw a line segment whose length is about 10 centimeters.

2-7. For your own body, referring to the drawing on page 52, what is the length of each? (Measure in centimeters.)

2. digit (thickness of index finger)   3. hand

4. cubit   5. foot

6. span   7. yard

8. Why are standardized measures useful?

9. The metric system was first established in what country and in what year?

10. In the early 1800's, why didn't the U.S. adopt the metric system?

11-13. Give the meaning of each prefix.

11. milli-   12. centi-   13. kilo-

14-27. Fill in the blanks.

14. The basic metric unit of distance is the _____.

15. The basic metric unit of mass or weight is the _____.

16. 1 millimeter = _____meters   17. 1 milligram = _____grams

18. 1 meter = _____ millimeters   19. 1 gram = _____milligrams

20. 1 kilogram = _____grams   21. 1 kilometer = _____meters

22. 1 centimeter = _____meters   23. 1 meter = _____centimeters

24. 1 kilogram ≈ _____pounds   25. 1 inch = _____centimeters

26. 1 meter ≈ _____inches   27. 1 mile ≈ _____kilometers

28. A meter is a little bigger than

(a) an inch   (b) a foot   (c) a yard   (d) a mile

29-34. Give the abbreviation for each unit.

29. meter   30. centimeter   31. kilometer

32. milligram   33. kilogram   34. gram

35-40. An abbreviation is given. Give the unit.

35. kg   36. cm   37. m   38. g   39. mm   40. km

Questions testing understanding of the reading

1. Why is a world-wide system of measurement more useful today than it was 100 years ago?

2. When the United States and Soviet Union make joint space efforts, are the measurements in the English system or the metric system?

3. A 110-pound student weighs about

(a) 5 kilograms   (b) 50 grams   (c) 50 kilograms

4. The height of a tall basketball player is about _____ meters.

5-10. International races in swimming or track are now always run in metric distances. It helps to know that 110 yards ≈ 100 meters.

5. 200 meters ≈ _____ yards

6. 400 meters ≈ _____ yards

7. 800 meters ≈ _____ yards

8. 1500 meters ≈ _____ yards

9. 1600 meters ≈ _____ yards

10. 3000 meters ≈ _____ yards

11-12. Large distances are measured in kilometers.

11. Would you be able to walk 1 kilometer?

12. Could a person drive 100 kilometers in a single day?

13-14. The metric unit of volume is the <u>liter</u>. (Many English units are used, including the quart, gallon, bushel, pint, teaspoon, and cup.) A liter is a little larger than a quart.

13. How many milliliters in a liter?

14. How many liters are in two kiloliters?

15. On every cereal box, the amount of protein per serving is listed. What metric unit measures this amount?

16. <u>Photography</u>. You have probably heard of 8mm, 16mm, and 35mm film. (a) Draw segments with these lengths. (b) What do these lengths stand for?

17. Draw a segment which has the length of a 100mm cigarette.

18-25. <u>Test your knowledge of the English system</u>. Most people cannot answer all of these questions correctly. (The English system is difficult to learn.)

18. How many feet in a yard?

19. How many yards in a mile?

20. How many ounces in a pound of feathers?

21. How many ounces in a pound of gold?

22. How many pounds in a short ton?

23. How many pounds in a long ton?

24. What is the abbreviation for "ounce"?

25. What is the abbreviation for "pound"?

1. Thomas Jefferson and Robert Morris helped to establish our
    decimal monetary system. Why was Thomas Jefferson so
    interested in how things are measured?

2. Why will tools made for U.S. cars not work for Volkswagens?

3. The meter is now defined as "1,650,763.73 wave lengths in
    a vacuum of the red-orange radiation emitted by the trans-
    ition between the energy levels $2p_{10}$ and $5d_5$ of the krypton-86

    atom." (You are not expected to understand the definition.)
    Why do you think such accuracy is needed?

4. Genesis 6:15 gives dimensions of Noah's Ark. "And this is
    how thou shalt make it: the length of the ark three hundred
    cubits, the breadth of it fifty cubits, and the height of it
    thirty cubits." The exact length of the cubit used by Noah
    is not known. It is estimated at between 17 and 21 inches.
    In feet, how long, high, and wide was Noah's ark?

Lesson 10

The Real Numbers and Measurement

Here are the names given to the kinds of numbers mentioned
in previous lessons. Begin with the natural numbers 1, 2, 3, ...
The natural numbers, their opposites, and zero make up the
integers.

the set of integers = $\{0, 1, -1, 2, -2, 3, -3, 4, -4, \ldots\}$

Counting problems--even those with two directions--seldom
need more than integers. But comparison of integers leads to
other numbers, the rational numbers.

58

Definition:

> A <u>rational number</u> is a number
> which can be the answer to a
> division problem using integers.

In Lesson 2, you learned that $\frac{2}{3}$, 30%, 1.25, and 659 are
examples of rational numbers. Their opposites $-\frac{2}{3}$, -30%, -1.25,
and -659 are also rational numbers.

Measurement problems lead to decimals.

Definition:

> A <u>real number</u> is a number which
> can be represented by a decimal.

A real number may be positive, negative, or zero. The
decimal may be ending or unending. Recall from Lesson 9 that
every rational number can be written as an ending or repeating
decimal. So every rational number is a real number.

But it is easy to make up a decimal which does not end or have
a repeating pattern.

.01002000300004000005000000600000007...

This decimal could not stand for a rational number. It stands for
a number which is called <u>irrational</u>.

Definition:

> An <u>irrational number</u> is a real number
> which is not a rational number.

59

This diagram shows how these numbers are related.

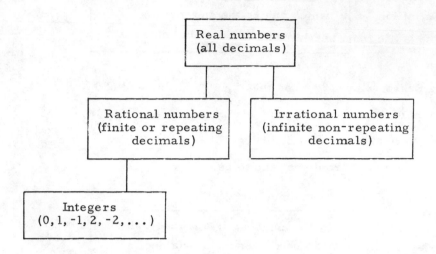

Each of these types of numbers includes both positive and negative numbers. The positive integers are the natural numbers.

As you know, lengths, temperatures, masses, etc., cannot be measured exactly. There are two reasons why. (1) No ruler (or other measuring device) can be perfectly accurate. The marks on it have thickness and under a microscope are not straight. (2) The length (or other thing) being measured changes with the wind, temperature, and pressure.

So a temperature of -5° can only be an approximation. You might only be able to say that the temperature is between -5.001° and -4.999°. However, it is <u>convenient</u> to use <u>one</u> number to stand for length or temperature. Sometimes the best number is a

60

rational number. Sometimes the best number is an irrational

number.

One of the most famous irrational numbers is called "pi,"

written π. Pi arises in the following way. Take a spool of

wire with a diameter of 1 unit.

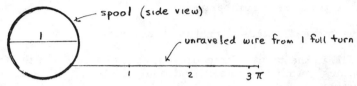

Now turn the spool one complete revolution and let the wire

unravel (pictured above). The length of wire unraveled is best

approximated by π. By fairly complicated methods, it is

possible to calculate pi to many decimal places. In some

applications, 10 or 12 or even more decimal places of π are

needed. Here are 20 decimal places.

π = 3.14159265358979323846...

Pi is known to be irrational. So its decimal does not repeat or end.

When rough estimates are enough, the approximations 3.14,

3.14159 or $\frac{22}{7}$ are used.

1. Name three integers.

2-4. Define:

2. rational number   3. real number   4. irrational number

5. Name three rational numbers.

6. Name three real numbers.

7. Give two examples of irrational numbers.

8-13. Fill in the blank with every correct choice from the following list: natural number, integer, rational number, irrational number, real number.

8. π is a(n) _____.   9. 5 is a(n) _____.

10. The decimal for a(n) _____ either repeats or ends.

11. The decimal for a(n) _____ is unending and does not repeat.

12. -3.24 is a negative _____.

13. 6.$\overline{48}$ is a positive _____.

14-19. Tell whether the number is rational or irrational.

14. -2.3   15. -2.$\overline{3}$ 16. -2.3030303030...

17. 612,000   18. π 19. 3.14159

20. Which numbers in Questions 14-19 are real numbers?

21. Which numbers in Questions 14-19 are integers?

22. True or False: When you use $\frac{22}{7}$ for π, you are using a rational approximation to an irrational number.

23. Some common approximations for π are _____, _____, and _____.

24. Why can't lengths be measured exactly?

25. Under a powerful microscope, will the edge of this paper look straight?

<u>Questions testing understanding of the reading</u>

1. This lesson contains an infinite decimal that begins .01002...
    Show what you think are the next 50 digits <u>after</u> those given
    on page 59.

2. A nurse says your temperature is 99.4°. It is probably
    more accurate to say that it is between _____ and _____.

3. You have a good scale which measures to the nearest .1 kg.
    If the scale shows a weight of 50.3 kg, your weight is probably
    between _____ and _____.

4-7. Give the first 10 decimal places in the decimal expansion of:

4. $\pi + 1$ 5. $\pi + \pi$ 6. $\pi/2$ 7. 100π

8. A person announces: "I have cut this wire so that it is
    exactly 2 cm long." What can you say about this?

9-13. Fill in the blank with <u>every</u> correct choice from the following
list: natural, rational, negative, irrational, real.

9. Every _____ number is a real number.

10. Every rational number is a _____ number.

11. Every irrational number is a _____ number.

12. Every _____ number is a natural number.

13. Every natural number is a _____ number.

14. Copy this diagram. Fill in each box with one of the words
    integer, rational, irrational, real.

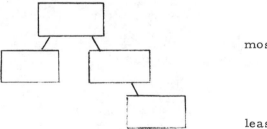

most inclusive

least inclusive

63

Questions for discussion and exploration

1. At one time scientists thought that the position of electrons might be <u>exactly</u> found. (This would possibly allow exact measurements.) That was until the Heisenberg Uncertainty Principle (1927), put forth by Werner Heisenberg, a German physicist. Using your school library, find out what this principle is.

2. Find a tin can in good condition. Measure the diameter of the can to the nearest millimeter. Mark the can and roll it once to measure the distance around the top of the can. Divide the distance around by the diameter to approximate π.

3. Around 480 A.D., a Chinese worker in mechanics, Tsu Ch'ung-chih, used $\frac{355}{113}$ as an approximation for π. (Europeans did not have this good an approximation until over 1000 years later.) To how many decimal places is the approximation accurate?

4. The words used to describe numbers usually have different meanings outside of mathematics. Give a non-mathematical meaning for each word.

   (a) natural   (b) rational   (c) irrational

   (d) positive   (e) negative   (f) real

The goal of this chapter is to show many applications of _real_ _numbers_. The reals are the numbers which can be written as finite or infinite decimals. Decimals are most convenient for measurement because they are easily compared, added, and multiplied.

Because various types of measurements are found so often, it is helpful to have a system of measurement which is based upon the decimal system. The _metric system_ is such a system. Its fundamental units for distance (meter) and mass (gram) are used by almost everyone in the world.

Some situations have two opposite directions in which one can measure. For these situations, _negative numbers_ are used.

Those numbers which can be represented by ending or repeating decimals are called _rational_. They also can be written as the quotient of two integers. This makes rational numbers particularly useful in comparison.

Comparisons often lead to percentages. These can be pictured in circle graphs. Measurements can be pictured in bar graphs.

A special type of comparison occurs in many experiments and in sampling. It is the <u>relative frequency</u> of an outcome and equals the quotient

$$\frac{\text{frequency of an outcome}}{\text{number of repetitions of an experiment}}.$$

Relative frequencies are always rational numbers between 0 and 1.

The most common numbers are the <u>natural</u> numbers 1, 2, 3, ..., used in counting, ordering, and identification. The <u>integers</u> include the natural numbers, their opposites -1, -2, -3, ..., and zero.

CHAPTER 2

PATTERNS AND VARIABLES

Lesson 1

Mathematics - The Study of Patterns

Scientists try to find patterns in the real world. These patterns are then applied by people who are not scientists. For example, a biologist may test fertilizers, looking for the following pattern:

Presence of fertilizer X leads to better corn growth.

When a biologist finds such a fertilizer, farmers will consider it for use. (Because agronomy-- the study of crops-- is quite advanced in the U.S., our crop yields are often the highest in the world.)

Here are other more general examples.

Scientist	looks for patterns in the behavior and properties of	applied to the work of
Sociologist	social groups	social workers, advertisers, governmental agencies, counselors...
Economist	financial matters	business, savers, investors, labor leaders...
Physicist	physical objects	inventors, tool makers, carpenters, plumbers, engineers...
Biologist	living things	ranchers, doctors, owners of pets, gardeners, farmers...

What is mathematics?

> Mathematics is the study of patterns.

Mathematicians study the patterns which they or others have found. Here are some of the branches of mathematics and the types of patterns studied.

	Branches*	Study patterns of or in*
	Algebra	structures, number patterns
Larger branches	Geometry	pictorial ideas
	Analysis	functions, limits
	Computer Science	algorithms, languages
	Probability	uncertainties
	Statistics	data
Smaller branches	Number Theory	natural numbers
	Logic	reasoning
	Topology	continuous deformations, networks

*Both the branches and the patterns they study are greatly oversimplified in this table. The branches often overlap. The patterns are often quite complex.

Because mathematics studies patterns, it is used daily by almost all scientists. And for this reason it is needed by those who apply the work of scientists.

68

There is a limitless variety of patterns.

Example 1: <u>An arithmetic pattern.</u> Three instances are given.

$$2 + 3 = 3 + 2 \qquad \frac{1}{11} + \frac{2}{3} = \frac{2}{3} + \frac{1}{11} \qquad 862.9 + 1 = 1 + 862.9$$

Description: You can add two real numbers in either order and get the same answer. This is such a common pattern that it has a special name, <u>commutativity of addition.</u>

Example 2: <u>A geometric pattern.</u> Two special cases are given.

Description: The sum of the measures of the angles of a triangle is 180°.

Example 3: <u>A pattern in driving.</u> If you drive at a rate of 50 mph, then in 1 hour you will travel 50 miles, in 2 hours you will travel 100 miles, and in 3 hours you will travel 150 miles.

Description: At 50 mph, the distance travelled in miles is 50 times the time spent in hours.

Example 4: <u>A pattern whose description is not true</u>.

Concerning the odd natural numbers:

3 has only the factors 1 and 3

5 has only the factors 1 and 5

7 has only the factors 1 and 7

Description: Every odd number has as its only factors 1 and itself. (The description is not true because the odd number 9 has 3 as a factor.)

69

These examples show the three things mathematicians do with patterns. Mathematicians try to:

1. Find patterns.

2. Describe patterns (English words often are too long.).

3. Determine the truth or falsity of patterns.

Questions covering the reading

1. _____ try to find patterns in the real world. _____ study these patterns.

2. Name three types of scientists and what they do.

3. Name three of the larger branches of mathematics and what they study.

4. Name three of the smaller branches of mathematics and what they study.

5-7. Give an example of a pattern which is

5. from arithmetic   6. from geometry   7. from driving

8. What are the three things mathematicians do with patterns?

9. As sociologists find more patterns in the behavior of social groups, their work becomes more mathematical. What people apply the work of sociologists?

10. What people apply the work of biologists?

Questions extending the reading

1-2. On p. 71 is a table of the 21 smallest natural numbers and their factors. The table has many patterns.

1. One pattern is: "1 is a factor of every natural number." Find another pattern.

Number	Factors	Number	Factors	Number	Factors
1	1	8	1, 2, 4, 8	15	1, 3, 5, 15
2	1, 2	9	1, 3, 9	16	1, 2, 4, 8, 16
3	1, 3	10	1, 2, 5, 10	17	1, 17
4	1, 2, 4	11	1, 11	18	1, 2, 3, 6, 9, 18
5	1, 5	12	1, 2, 3, 4, 6, 12	19	1, 19
6	1, 2, 3, 6	13	1, 13	20	1, 2, 4, 5, 10, 20
7	1, 7	14	1, 2, 7, 14	21	1, 3, 7, 21

2. A student answers Question 1 with the following pattern: Every natural number has less than 7 factors. Why is this pattern incorrect?

3. Find a pattern in this table of library fines.

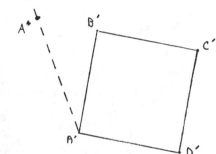

Number of Days Overdue	Cost
1	.10
2	.15
3	.20
4	.25
5	.30
6	.35
7	.40
8	.45
9	.50

4. Trace the two squares at right. Find the point exactly halfway between A and A'. Call it A*. (This is done for you.) Do the same for B and B', calling the point B*. Find C* and D*. Now connect A*, B*, C*, and D*. Is there any pattern?

5. The decimal for 1/98 begins .01020408... and doesn't repeat until over 80 decimal places. Find a pattern in this decimal and guess what the next 10 places are.

6. Examine the following true statements. (The raised dots stand for multiplication.)

$$3 \cdot 3 - 3 = 2 \cdot 2 + 2$$
$$80 \cdot 80 - 80 = 79 \cdot 79 + 79$$
$$11 \cdot 11 - 11 = 10 \cdot 10 + 10$$

(This problem continues on page 72.)

(a) Find a pattern. Use it to get 3 more statements like those given.

(b) Describe your pattern in English words.

(c) Do you think the pattern works for fractions?

7. <u>A misleading pattern.</u> Examine the following true statements.

$$1 - \frac{1}{2} = 1 \cdot \frac{1}{2}$$

$$5 - \frac{5}{6} = 5 \cdot \frac{5}{6}$$

$$\frac{1}{2} - \frac{1}{3} = \frac{1}{2} \cdot \frac{1}{3}$$

A person describes the pattern as follows: Subtraction is the same as multiplication. What is wrong with that description?

8. There is a pattern connecting area, length, and width of rectangles. What is this pattern?

9. Examine the following data. (source: Passport Office, U.S. State Department)

Year	1965	1966	1967
No. of passports issued	1,330,290	1,547,725	1,685,512

1968	1969	1970	1971
1,748,416	1,820,192	2,219,159	2,398,968

(a) Find a very simple pattern in this data.

(b) Name some groups of individuals who might be interested in this pattern.

10. There is a pattern used in multiplying fractions. In English, describe this pattern.

<u>Questions for discussion and opinion</u>

1. Meteorologists try to find patterns in weather. So their work is used by everyone. What mathematics must you know to understand the weather?

2. Why might a musician be interested in the work of physicists?

Lesson 2

Words and Symbols of Arithmetic

When a pattern is difficult to describe in English, we try to find an easier way to describe it. If the pattern occurs again and again, mathematicians will invent symbols or words which stand for English words or phrases.

Examples of symbol invention

1. Fractions like $\frac{2}{3}$ and $\frac{10}{5}$ first appeared about the same time as decimals (800-1200 A.D.). It is not known who first used the bar symbol for fractions.

2. Symbols for operations were invented later.

Symbol	Inventor	Date of invention
+ (addition)	Johann Widman	1489
- (subtraction)	Johann Widman	1489
= (equals)	Robert Recorde	1557
<, > (inequality)	Thomas Harriot	1631
X (multiplication)	William Oughtred	1631
÷ (division)	John Pell	1631

Before these dates the operations were done. However, words were used instead of symbols.

3. Suppose you change 3. 7 to 4

    61. 39 to 61

    -2. 1 to -2

    This process happens so often it is given a special name:

    rounding to the nearest integer.

4. Answers to the common operations each have special names.

$2 + 3$	sum of 2 and 3
$8 \cdot \frac{6}{5}$	product of 8 and $\frac{6}{5}$
$19 - 4.6$	19 less 4.6, 4.6 from 19
$\frac{3}{100}$ or $3 \div 100$	3 divided by 100

    The answer to a subtraction problem is called a difference.

    The answer to a division problem is a quotient.

    In arithmetic, Oughtred's symbol $\times$ is used for multiplication. In algebra, a raised dot · is used. From here on in this course, you should use the dot for multiplication. You should not use $\times$ for multiplication. (The $\times$ would become confused with uses of the letter x.)

    When the symbols for operations are used, confusion can arise. For example, starting with $50, spending $10, and then spending $6, you will have left

    50 - 10 - 6 dollars.

As you know, this is $34.

74

But someone else, seeing 50 - 10 - 6, might calculate

10 - 6 = 4 and so gets $46 as a wrong answer.

To avoid confusion, language rules are needed. These rules

tell which operations should be done first. They are agreed upon

world-wide. (In this way, written mathematics is like written

music. Musicians from most countries can read the same music.)

Language Rules for Order of Operations

In any expression, operations are done in the following order.

First: Work inside parentheses, from the inside out.

Then: Do multiplications or divisions from left to right.

Last: Do additions or subtractions, left to right.

Examples: Order of operations

Explanations:

1. 50 - 10 - 6

= 40 - 6

= 34

There are only subtractions so work from left to right.

2. 9 + 10 · 2

= 9 + 20

= 29

Do multiplications or divisions before additions or subtractions.

3. 50 - (10 - 6)

= 50 - 4

= 46

This is different from Example 1. Work in parentheses first.

4. $100 - (40 + 9 \cdot 2) \div 2$ First work inside paren-
                                             theses. Multiplication
   $= 100 - (40 + 18) \div 2$ is first there.

   $= 100 - 58 \div 2$ Now division before sub-
                                             traction.
   $= 100 - 29$

   $= 71$

The fraction bar is itself a parentheses symbol. That is,

$$\frac{12 + 3 \cdot 8}{4 + 2} \quad \text{is short for} \quad \frac{(12 + 3 \cdot 8)}{(4 + 2)}$$

So you must work in the numerator and denominator before dividing.

5. $\dfrac{(12 + 3 \cdot 8)}{(4 + 2)} = \dfrac{(12 + 24)}{6} = \dfrac{36}{6} = 6$

Questions covering the reading

1. Why do mathematicians invent symbols or words?

2. Give some examples of words which have been invented by
   mathematicians.

3. Multiple choice: The symbols +, -, ·, and ÷ are:
   (a) less than 500 years old.
   (b) between 500 and 1000 years old.
   (c) between 1000 and 2000 years old.
   (d) over 2000 years old.

4. What did mathematicians do before the symbols in Question 3
   were invented?

5. In algebra, what symbol is used for multiplication?

6-9. What name is given to the answer to:

6. an addition problem.   7. a subtraction problem.

8. a multiplication problem. 9. a division problem.

76

10-13. Calculate:

10. the product of 4 and $\frac{2}{3}$ 11. the sum of $\frac{8}{5}$ and $\frac{1}{5}$

12. 40 less 22   13. 2 divided by 6

14. Give an example to show that rules for order of operations are necessary.

15. What are the rules for order of operations?

16-30. Simplify:

16. $14 - 3 \cdot 4$ 17. $8 \div 4 - 2$

18. $100 - (80 - 10)$ 19. $12 \div 6 \div 2$

20. $13 \cdot (4 - 1)$ 21. $2 \cdot 10 - 5$

22. $\frac{9 - 6}{3}$ 23. $\frac{3 + 2}{3 - 2}$

24. $\frac{1}{2} + \frac{1}{2 + 2}$ 25. $2 \cdot 5 + 3 \cdot 3$

26. $4 + 3 \cdot 6 + 2$ 27. $11 - 2 \cdot 3 + 4$

28. $12 \div 3 + 3 \div 2$ 29. $\frac{100 + 10}{10}$

30. $\frac{14 + 6}{2 + 3}$

Questions testing understanding of the reading

1-2. Multiple Choice.

1. The sum of 40 and the product of 2 and 5 is

(a) $(40 + 2) \cdot 5$ (b) $40 \cdot (2 \cdot 5)$ (c) $40 + 2 \cdot 5$

2. The product of $\frac{1}{2}$ and $\frac{2}{3}$ less the sum of $\frac{3}{4}$ and $\frac{1}{5}$ is

(a) $\frac{1}{2} + \frac{2}{3} - \frac{3}{4} \cdot \frac{1}{5}$ (b) $\frac{1}{2} \cdot \frac{2}{3} - \frac{3}{4} + \frac{1}{5}$

(c) $\frac{1}{2} \cdot \frac{2}{3} - (\frac{3}{4} + \frac{1}{5})$ (d) $\frac{1}{2} + \frac{2}{3} - (\frac{3}{4} + \frac{1}{5})$

3-6. True or False.

3. $13 + (4 \cdot 2) = 13 + 4 \cdot 2$ 4. $18 - (5 - 2) = 18 - 5 - 2$

5. $\dfrac{12 + 6}{18 + 24} = \dfrac{(12 + 6)}{(18 + 24)}$ 6. $2 + 3 \cdot 4 - 5 = (2 + 3) \cdot 4 - 5$

C   7-12. Round to the nearest integer.

7. $1.3 \cdot 2.4 + 3.0$ 8. $16.2 - 2.4 \cdot 3.9$

9. $9 \cdot 9 \cdot 9 \cdot 2 \cdot 3 \cdot 0$ 10. $88 \cdot 125 - 100 \cdot 3$

11. $6.3 \div (2.09 + 3.12)$ 12. $1 \div 6 \div 7$

Questions for discussion

1. The raised dot symbol for multiplication was used by Harriot in the early 1600's. It was not popular until 1698 when Gottfried Leibniz adopted it. Why would it take so long for a symbol to become popular? Why was Leibniz so influential?

2. Why would it be natural for symbols for numbers to be invented before symbols for operations?

3. Why is it helpful to have worldwide agreement on language rules?

4. The language of mathematics is worldwide. So is the language of music. Are there other worldwide languages?

5. Do more people in the world speak English than any other language?

Skill review.   (Answers upside down at end of lesson.)

1-10. Round to the nearest integer.

1. 2.8   2. 461,312.29   3. $\dfrac{15}{7}$ 4. $-\dfrac{2}{3}$ 5. 11

6. $\dfrac{1}{3.5}$ 7. 7.7731204   8. -12   9. $-6\dfrac{3}{7}$ 10. 18.39

11-20. Round to the nearest hundred.

11. 346   12. 4.3801   13. -61

14. 8305   15. $\dfrac{4000}{3}$ 16. $100 \cdot \pi$

78

17. $\dfrac{862}{2}$ 18. 9949.99   19. $\dfrac{1}{236}$

20. -48394

Lesson 3

Describing Patterns Using Variables

Here are 3 instances of a pattern.

$$19 + 0 = 19 \qquad 1.6 + 0 = 1.6 \qquad \dfrac{3}{2} + 0 = \dfrac{3}{2}$$

The pattern can be described in words. "If zero is added to a real number, the sum is that number."

That description has two weaknesses. It is long. It doesn't look like the pattern. A shorter and nicer-looking description uses a single letter to stand for the numbers which change in the instances.

If r is a real number, $r + 0 = r$

The letter r in this description is called a <u>variable</u>. In this pattern, r could be replaced by any real number. Above r was replaced first by 19, then by 1.6, then by $\dfrac{3}{2}$.

Definition:

In the above example, r could be any real number. So the set of replacements for r is the set of real numbers. If r is replaced by 15 we write r = 15. Then the instance of the pattern is 15 + 0 = 15.

By using variables, infinitely many instances of the pattern are described at the same time. For this reason:

<u>Variables are the most common</u>

<u>symbols used in the description</u>

<u>of patterns.</u>

Examples: Describing Patterns Using Variables

1. Here are 3 instances of a pattern. You should check the arithmetic.

$$\frac{50 \cdot 3}{3} = 50$$

$$\frac{\frac{1}{2} \cdot 3}{3} = \frac{1}{2}$$

$$\frac{9.2 \cdot 3}{3} = 9.2$$

<u>Description using variables:</u> Let v be any real number.
Then $$\frac{v \cdot 3}{3} = v$$

Description in words: If a number is multiplied by 3 and the product is divided by 3, then the result is the original number. (The description using variables is much shorter.)

2. If 3 numbers appear in a pattern, then 3 variables may be needed.

$$2 \cdot 5 \cdot 3 = 3 \cdot 5 \cdot 2$$

$$\frac{8}{3} \cdot \frac{2}{9} \cdot \frac{1}{8} = \frac{1}{8} \cdot \frac{2}{9} \cdot \frac{8}{3}$$

$$0 \cdot 9 \cdot 7 = 7 \cdot 9 \cdot 0$$

Description: Let x, y, and z be any real numbers. Then

$$x \cdot y \cdot z = z \cdot y \cdot x$$

Because x is so often used as a variable, it is _never_ used to indicate multiplication.

3. Here is a pattern. You should check the additions and multiplications.

$$6 + 1.2 = 6 \cdot 1.2$$

$$2 + 2 = 2 \cdot 2$$

$$\frac{4}{3} + 4 = \frac{4}{3} \cdot 4$$

Description: Let a and b be any positive numbers. Then

$$a + b = a \cdot b$$

This description is correct for these instances but the pattern is not true in general. If a = 2 and b = 3, then we have 2 + 3 = 2 · 3. Since 5 ≠ 6, the pattern is not always true.

81

It is not always easy to tell whether a pattern is true. The best way is to substitute numbers for the variables.

Questions covering the reading

1. Define: variable.

2-5. Give 3 instances of each pattern.

2. If t is a positive number, then $\frac{t}{t} = 1$.

3. Let a and b be any numbers. Then $a + b = b + a$.

4. $v \cdot 8 \div 8 = v$.   5. $x \cdot y \cdot z = z \cdot y \cdot x$

6. Is it possible to have a pattern which is only sometimes true?

7. Why is the symbol x not used in algebra for multiplication?

8. Why are variables better than words in describing patterns?

Questions testing understanding of the reading

1-6. Give 3 instances for each pattern.

1. If x is any real number, then $\frac{2 \cdot x}{2} = x$.

2. If a and b are any real numbers, then $a \cdot b = b \cdot a$.

3. Let y stand for any prime number. Then y has as its only factors 1 and y.

4. Suppose z is a positive real number and z is not an integer. Then $z + 1$ is not an integer.

5. If m is any real number, then $7 \cdot m - 6 \cdot m = m$.

6. If t is any real number, then $\frac{1}{2} \cdot t + \frac{1}{2} \cdot t = t$.

7-10. Some instances of a pattern are given. Describe the possible general pattern. Only one variable is needed.

7. $6 \cdot 1 = 6$ 8. $\frac{34}{34} = 1$ $\frac{1.6}{1.6} = 1$
      $14.3 \cdot 1 = 14.3$
        $\pi \cdot 1 = \pi$
        $0 \cdot 1 = 0$ $\frac{7}{7} = 1$

9.
$$7 + 7 = 2 \cdot 7$$
$$11.3 + 11.3 = 2 \cdot 11.3$$
$$37 + 37 = 2 \cdot 37$$
$$66 + 66 = 2 \cdot 66$$

10.
$$4712 - 4712 = 0$$
$$\frac{1}{6} - \frac{1}{6} = 0$$
$$4 - 4 = 0$$
$$.031 - .031 = 0$$

11-18. Follow the directions of Questions 7-10. Two or more variables are needed in the description.

11.
$$5 \cdot 14 = 14 \cdot 5$$
$$47.2 \cdot 31.6 = 31.6 \cdot 47.2$$
$$\frac{1}{2} \cdot 100 = 100 \cdot \frac{1}{2}$$

12.
$$6 \cdot 3 + 6 \cdot 4 = 6 \cdot (3 + 4)$$
$$6 \cdot 11 + 6 \cdot \frac{1}{3} = 6 \cdot (11 + \frac{1}{3})$$
$$6 \cdot 7 + 6 \cdot 100 = 6 \cdot (7 + 100)$$

13.
$$\frac{5}{3} - \frac{1}{3} = \frac{5 - 1}{3}$$
$$\frac{18}{14} - \frac{11}{14} = \frac{18 - 11}{14}$$
$$\frac{9}{130} - \frac{6}{130} = \frac{9 - 6}{130}$$

14.
$$\frac{1 \cdot 3}{2 \cdot 3} = \frac{1}{2}$$
$$\frac{46 \cdot 3}{3 \cdot 3} = \frac{46}{3}$$
$$\frac{11 \cdot 3}{10 \cdot 3} = \frac{11}{10}$$

15.
1 tape costs $1 \cdot 4.98$
2 tapes cost $2 \cdot 4.98$
13 tapes cost $13 \cdot 4.98$
6 tapes cost $6 \cdot 4.98$

16. In 2 years, there will be $2 \cdot 500$ more people in the town.
In 14 years, there will be $14 \cdot 500$ more people in the town.
In 5 years, there will be $5 \cdot 500$ more people in the town.

17.
One cow has $4 \cdot 1$ legs.
Six cows have $4 \cdot 6$ legs.
80 cows have $4 \cdot 80$ legs.

18.
$$100m \approx 110 \text{ yd}$$
$$2 \cdot 100m \approx 2 \cdot 110 \text{ yd}$$
$$14 \cdot 100m \approx 14 \cdot 110 \text{ yd}$$

19-20. Given are 4 instances of a pattern. The statements are true.
(a) Give what seems to be the obvious description of this pattern.
(b) Show that your description is not always true.

19.
$$5 \cdot 2 < 100$$
$$16 \cdot 2 < 100$$
$$27 \cdot 2 < 100$$
$$7\frac{1}{2} \cdot 2 < 100$$

20.
$$17 \cdot 17 > 17$$
$$3 \cdot 3 > 3$$
$$105 \cdot 105 > 105$$
$$2.9 \cdot 2.9 > 2.9$$

Lesson 4

Words and Symbols of Algebra

The Greek mathematician Diophantus in about 250 A.D. used a single letter to stand for a number. But his work was lost for a long time. In the 1580's the mathematician Francois Vieta read the work of Diophantus. Vieta seems to have invented the idea of a variable--a letter which can be replaced by different numbers.

Consider the expression

$$40 \cdot n + 38$$

When n is replaced by 3, $40 \cdot n + 38 = 40 \cdot 3 + 38 = 158$. We call 3 a <u>value</u> of the variable n. A value of a variable is a number which is substituted for the variable. 158 is the corresponding value of the expression. You could write: When $n = 3$, $40 \cdot n + 38 = 158$.

Examples: Values of expressions

Expression	Value of variable	Value of expression
1. $(n+3)(n+2)$	5	$(5 + 3) \cdot (5 + 2) = 8 \cdot 7 = 56$
	8	$(8 + 3) \cdot (8 + 2) = 11 \cdot 10 = 110$
2. $12 \cdot x + 3$	$\frac{1}{2}$	$12 \cdot \frac{1}{2} + 3 = 6 + 3 = 9$
	4	$12 \cdot 4 + 3 = 48 + 3 = 51$
3. $18 - 8 \cdot t - t$	1	$18 - 8 \cdot 1 - 1 = 9$
	2	$18 - 8 \cdot 2 - 2 = 0$

84

In Examples 1-3, a dot was used to indicate multiplication. This symbol is usually deleted when variables are being multiplied.

With · for multiplication	Without ·
6 · y	6y
(n + 3)·(n + 2)	(n + 3)(n + 2)
18 - 8·t - t	18 - 8t - t
4·(x + 3)	4(x + 3)

Whether or not there is a symbol for the multiplication, the rules for order of operations still apply.

An expression may contain more than one variable. Then a value for each variable may be given.

Examples: Values of Expressions (continued)

4. The depth of a gear tooth is found by dividing the difference of the major and minor diameters by 2. Using variables, the depth is given by the expression $\frac{D - d}{2}$.

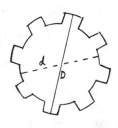

You must subtract before dividing because of the unwritten parentheses $\frac{(D-d)}{2}$.

If $D = 5.6$ cm and $d = 5$ cm, then the depth equals

$$\frac{5.6 - 5}{2} = \frac{.6}{2} = .3 \text{ cm}$$

85

5. Perhaps you begin saving for Christmas on October 1st.
Suppose you deposit $25 to begin. Then, adding d dollars
a week--there are 9 weeks--you will have

$$(25 + 9d) \text{ dollars}$$

by December 1st. The lack of a symbol between 9 and d
means they are multiplied. If you add 10 dollars a week, then
d = 10. So you will have

$$25 + 9 \cdot 10 \quad \text{or} \quad 115 \text{ dollars.}$$

Questions covering the reading

1-2. Who and when:
1. first used a single letter to stand for a number.
2. seems to have invented the idea of a variable.

3. (Review) Define: variable.

4. What is a <u>value</u> of a variable?

5-9. Give the value of the expression $15 + x$ when the value
of x is:

5. 10   6. 1/2   7. 0   8. 432   9. $1\frac{1}{3}$

10. Do the rules for order of operations apply to variables?

11. When variables are multiplied, what symbol is often not
found?

12. <u>Multiple Choice</u>. If m is replaced by 6 and x is replaced
by 4, then mx is equal to:

   (a) 10   (b) 24   (c) 64   (d) none of these

13-18. If d is replaced by 8, give the value of each expression.

13. 3d   14. d + 3   15. 2d - 5
16. 11.5 - d   17. 100 - 4d   18. $d + \frac{d}{2}$

19-28. If a is 10 and y is 5, give the value of:

19. the sum of a and y. 20. y divided by a.

21. the product of a and y. 22. ay + a

23. $\dfrac{a}{2a}$ 24. a less y

25. (3a)(y) 26. (a + y) + (a - y)

27. (a + y)(a - y) 28. $\dfrac{a + y}{a - y}$

29-30. Refer to Example 4. Find the depth of a gear tooth with:

29. major diameter 4 cm and minor diameter 3 cm.

30. major diameter 50 and minor diameter 32.

31-34. Refer to Example 5. How much will be saved by December 1st if each week:

31. $5 is deposited. 32. $7.50 is deposited.
33. $10 is deposited. 34. $12.50 is deposited.

Questions testing understanding of the reading

1-10. If a = 2, b = 3, c = 4, and d = 5, give the value of each expression.

1. b - a 2. bc 3. bc + cd

4. cb 5. $\dfrac{c + a}{c} + d$ 6. d - (b - a)

7. d - b - a 8. the product of d and a

9. b from ac 10. d divided by the sum of a and b

11-16. In the computer languages BASIC and FORTRAN, b times c is written as b*c. Here are expressions which might be found in a computer program in either of those languages. Evaluate:

11. 4 * 3 - 2 12. x * x when x = 7

13. 10 * (9 - 6.9) 14. 3 + 2 * 5 + 6

15. 18 - 4 * 2 16. 1 + 2 * 3 + 4 * 5 * 0

17-18. Phonograph records are shaped like flat cylinders. A 12" LP record gets its name because its diameter is about 12". Its thickness is about 1/16". The volume of a cylinder with diameter d and thickness t is 1/4 ddtπ. Find the amount of plastic in one 12" LP:

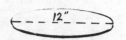

17. using the approximation $\frac{22}{7}$ for π.

C 18. using the approximation 3.14 for π.

19-22. Suppose you have d dollars. You spend t dollars. Then you spend s more dollars. You will then have left d - t - s dollars. Evaluate (find the value of) d - t - s when

19. d = 100, t = 40, s = 30

20. d = 100, t = 40, s = 50

C 21. d = 61.39, t = 28.42, s = 31.42

C 22. d = 197.61, t = 55.82, s = 131.94

23-28. Find the value of the expression when the value of the variable is (a) 0, (b) 1, (c) 2, (d) 3, (e) 4, (f) 5, and (g) 6. In each question, the answers you get should form a pattern.

23. 2n

24. 2m + 1

25. 10x + 9

26. $\frac{v}{v + 1}$

27. 7r + 3r

28. 11 - 2y (Guess on part (g).)

88

Lesson 5

Variables in Formulas

This rectangle has length 6 units and width 3 units. Its area may be found by counting the square units.

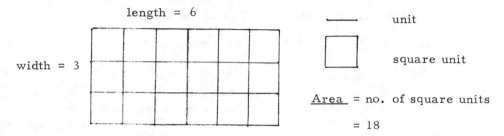

As you know, there is a pattern which relates length, width, and area.

The area of a rectangle equals its length times its width.

Using variables A, ℓ, and w to stand for the area, length, and width of this rectangle:

$$A = \ell \cdot w$$

This sentence gives a way to find the area of a rectangle. It is called a formula for the area of a rectangle in terms of its length and width.

Examples of Formulas:

1. Rate (in kilometers per hour) equals distance (in km) divided by time (in hours):

$$r = \frac{d}{t}$$

89

2. The perimeter of a triangle is the sum of the lengths of the sides of the triangle (all measured in same units):

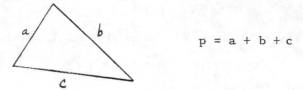

$$p = a + b + c$$

3. Here is a formula which gives profit in terms of selling price and dealer's cost (all measured in same units):

$$p = s - c$$

4. A conversion formula: L_I, a length in inches, is approximately equal to 39.37 times L_M, the length in meters.

$$L_I \approx 39.37 L_M$$

In formulas, the <u>units</u> used must be consistent. If you measure sides of a rectangle in inches and area in square feet, the formula $A = \ell w$ will not apply.

Letters used in formulas are chosen carefully. They often are the first letter of the thing they represent.

A for area

d for distance

In the same formula, capital and small letters almost always represent <u>different</u> things. The area of the ring at right is

$$\pi (R + r)(R - r).$$

R stands for the large radius

r stands for the small radius

In formulas, you should not change capital letters to small letters. Nor should you change small letters to capital letters. You may find it convenient to put curves in small letters. This helps distinguish them from capital letters. Here are the most confusing letters.

C F I J K L M S U V W X Y Z
c f i j k l m s u v w x y z

Questions covering the reading

1. The formula $A = lw$ gives the _____ of a rectangle in terms of its _____ and width.

2. What is the area (in square feet) of the floor of a room which is rectangularly shaped and 9 feet by 12 feet?

3. A room is 108 inches by 144 inches. What is its area in square feet?

4. What does Question 3 tell you about units in formulas?

5. In the formula $A = \pi(R + r)(R - r)$, do R and r stand for the same thing?

6. In a formula, can you change a small letter to a capital letter?

7. Write the alphabet from A to Z in capital letters. Then write the alphabet from a to z in small letters. Make sure your small letters look different (not just smaller) than your capital letters.

8-11. Consider the formula $p = s - c$, as in Example 3.

8. What do p, s, and c stand for?

9. What about letters in formulas makes Question 8 easy to answer?

10. This formula gives p in terms of _____.

11. Calculate p if s = $1 and c = 69¢.

91

1-6. Translate each sentence into a mathematical formula.

1. The total cost T of a number of similar articles is the number n of articles multiplied by c, the cost per article.

2. I, the amount of current (in amps) in a circuit, is the voltage V (in volts) divided by the resistance R (in ohms).

3. The perimeter P of a rectangle is twice the sum of its length ℓ and width w.

4. The area A of a triangle is half the product of its base b and its height h.

5. To find Celsius temperature C, subtract 32 from the Fahrenheit temperature F, then multiply the difference by 5/9.

6. The volume of a box is the product of its height, length, and width. (Use convenient letters.)

7-9. The safe working strength (in pounds) of a leather belt is approximately 300 times the product of its width and its thickness (in inches). (See also Question 1 on p. 94.)

7. Translate the sentence into mathematics. Use appropriate letters.

8. What is the safe working strength of a belt 2 inches wide and 1/8 inch thick?

9. Give the safe working strength of a belt 15 inches wide and 1/10 inch thick.

10-11. To estimate the number of bricks needed in a wall, some bricklayers use the formula

$$N = 7\ell h$$

where ℓ and h are the length and height of the wall (in feet).

10. How many bricks would this bricklayer need for a wall 8 feet high and 12 feet long?

11. A bricklayer has 2000 bricks. Could she build a wall 16 feet high and 20 feet long?

12-13. The area of a trapezoid with height h and bases b and B is given by the formula

$$A = \frac{1}{2}h(B + b)$$

Draw an accurate picture and calculate the area A when:

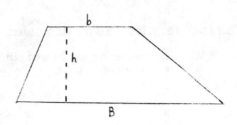

12. h = 4 cm, b = 6 cm, and B = 10 cm.

13. h = 3.1 cm, b = 2.0 cm and B = 3.4 cm.

14-17. The percentage of discount on an item is given by the formula

$$p = 100 \left(1 - \frac{n}{g}\right)$$

where g is the original price and n is the new price.

14. Find the percentage of discount on an item which is reduced from $20 to $17.

C 15. Find the approximate percentage of discount on an item which is reduced from $19.95 to $16.95.

16. Find the percentage of discount on an item which is reduced from $200 to $100.

17. Why is g used for the original price instead of o or p?

18. (Use the formula for the area of a ring given on page 90.) The owner wants to surround his circular garden by a walkway 1 meter wide. The garden has a radius of 10 meters. How much ground will he have to clear?

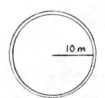

19. One basketball costs $7.98. Write a formula for the cost c in terms of the number of basketballs purchased.

20-21. When Ann was 12, Betsy was 15. Now Ann is 14 and Betsy is 17. Let A be Ann's age and B be Betsy's age.

20. Write a formula giving Ann's age in terms of Betsy's age.

21. Write a formula giving Betsy's age in terms of Ann's age.

22. Use Question 3, p. 92. Find the length of fence needed for a rectangular yard which is 30 feet by 10 feet.

Question for discussion and opinion

1. What is meant by "safe working strength" of a leather belt? Why would anyone care?

Lesson 6

Replacement Sets for Variables

In the area formula $A = \ell w$, ℓ and w stand for lengths of sides. So ℓ and w may each be replaced by any positive real number. The set of positive real numbers is called the <u>replacement set</u> for ℓ and w.

The sentence "c cows have 4c legs" only makes sense when c is 0 or a natural number. So the replacement set for c is $\{0, 1, 2, 3, 4, 5, \ldots\}$. That is, you can only replace c by one of the elements of that set. (For example, it would be silly to replace c by $6\frac{2}{5}$.)

Definition:
> <u>The replacement set of a variable</u> is the set of possible values of the variable.

Generally, the replacement set contains all values which could possibly happen. So you can often determine a replacement set by looking at how the numbers are used. Here are some of the uses of numbers and the usual replacement sets.

Use	Replacement Set
1. counting	$\{0, 1, 2, 3, 4, 5, \ldots\}$ = set of non-negative integers
2. measuring	set of positive real numbers
3. ordering	$\{1, 2, 3, 4, 5, \ldots\}$
4. situations with two opposite directions	set of all integers (if counting) set of all real numbers (if measuring)
5. relative frequency	set of rational numbers between 0 and 1, inclusive

Examples: Replacement Sets

1. Suppose p is the relative frequency of getting heads in tossing a coin. Then the replacement set for p is the set of numbers between 0 and 1. That is, it is possible that p could be 0, or 1, or $\frac{1}{2}$, or $\frac{2}{9}$ or any other rational number between 0 and 1.

2. Suppose P stands for the profit of a company. Profit is a situation with two possible directions. So P might be any real number. The replacement set for P is the set of all real numbers. (If P is negative, it would stand for a loss.)

3. Let r be the number of questions answered correctly on a

10-question test. Then the replacement set for r is

$\{0, 1, 2, 3, 4, 5, 6, 7, 8, 9, 10\}$.

Questions covering the reading

1. Define: replacement set of a variable.

2. Suppose the replacement set for v is $\{1, 6, 3\}$. Then what
    are the possible values of v?

3-8. Give the most suitable replacement set for a variable in
each situation.

3. measuring   4. counting

5. measuring situation with two opposite directions

6. counting situation with two opposite directions

7. relative frequency   8. ordering

9-16. Pick the most appropriate replacement set for the variable
in the given situation. Make your choices from the following list.

    set of all real numbers   set of positive real numbers
    set of all integers   set of nonnegative integers
    set of rational numbers between 0 and 1

9.   s, the number of students in a high school

10. P, the profit of a company

11. e, the relative frequency of heads in 1000 tosses of a coin

12. h, the height of a doorway

13. y, the number of yards gained by a runner in a football game

14. t, the finish of a horse in a horse race

15. c, the change in temperature from one day to the next

16. r, the number of refrigerators sold to Eskimos

1-2. Give a reasonable replacement set for each variable in the given formula. Pick your answer from the list given in Questions 9-16 p. 96.

1. C = 7.98n, where C is the cost of buying n basketballs.

2. A = ℓw, where A is the area of a rectangle with length ℓ and width w.

3. If the replacement set for v is $\{1, 2, 3\}$, is it true that v + 18 is always greater than 20?

4. If the replacement set for x is $\{2, 5, 7\frac{1}{2}, 11.5\}$, is it true that twice x is always an integer?

5. Let a have replacement set $\{1, 1\frac{1}{2}, 2\}$ and b have replacement set $\{4, 7\frac{1}{2}, 10\}$, is it true that a + b is always an integer?

6. Let p and q be elements of the set $\{0, 1, 2\}$. Then what values are possible for p + q?

7. If the replacement set for x is $\{$Benjamin Franklin, George Washington, John Adams$\}$, is it always true that x is the name of a President of the United States?

8. If the replacement set for x is the set of positive integers and the replacement set for y is the set of positive integers, is (x + y) always a positive integer?

9-12. x has replacement set $\{8, 14, \pi, 1\frac{1}{2}, 3.9\}$. Give all possible values of:

9. 3x   10. x + 3   11. $\dfrac{12x - 5}{10}$ 12. 100 - 2x

13-16. y has replacement set $\{0, 100, 200\}$. Give all possible values of:

13. (y + 1)(y + 2)   14. 2y + 3y + 4y

15. (y + 1)y + 2y + 2   16. 500 - 2y

97

1.   Suppose F stands for temperature in degrees Fahrenheit.
      Then the replacement set for F contains all positive real
      numbers and <u>some</u> negative numbers.   Why doesn't the
      replacement set contain all negative real numbers?

Lesson 7

Open Sentences and Solutions

An <u>open sentence</u> is a sentence which contains one or more
variables.   Here is an example.

There are s students in my algebra class.

This sentence concerns counting.   So a suitable replacement set
for s is the set of all nonnegative integers.   A value of s which
makes the statement true is a <u>solution</u> to the sentence.   For the
author's 1975-76 algebra class, the solution was 26.   There were
26 students on his class list.

Definition:
> Suppose an open sentence contains one
> variable.   A <u>solution</u> to the sentence is
> a replacement for the variable which
> makes the sentence true.

Some open sentences have many solutions. For instance, consider the sentence:

X has been President of the United States.

A suitable replacement set for X would be the set of all people who have been U.S. citizens. Of the hundreds of millions of replacements as of 1975, 37 are solutions to this sentence. Some of the solutions are

George Washington, John Kennedy, Millard Fillmore,...

Some replacements which are not solutions are

Martin Luther King Jr. , Eleanor Roosevelt, Johnny Carson,...

If a variable stands for a number, when a replacement set is not given, assume that it is the set of all real numbers. Examples:

1. Open sentence: $x + 3 = 10$

Assumed replacement set: set of all real numbers

Solution: 7

2. Open sentence: $5 > y$

   Solutions: Any number less than 5 will work. Thus there are many many solutions. Some are -100, .67, 2, π, 4.999. The solutions cannot all be listed but they can be graphed. (The circle around the point for 5 shows that 5 itself is not a solution.)

   Graph of solutions:

Examples 3 and 4 below involve similar open sentences. One comes from a counting problem. One comes from a measuring problem. So the replacement sets are different. This affects the solutions.

3. That board is between 38 and 43 cm long. Its length is x.

   Replacement set: set of positive real numbers

   Sentence: $38 \leqslant x \leqslant 43$

   Graph of all solutions:

4. There are from 38 to 43 students out for football this year. The number of students out is n.

   Replacement set: set of non-negative integers

   Sentence: $38 \leqslant n \leqslant 43$

   Graph of all solutions:

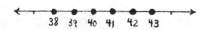

There are open sentences in many areas other than mathematics. But variables are not always used. Instead of variables, blanks or question marks may be used. These are also open sentences.

100

_____ is in Baseball's Hall of Fame.

The country with the most oil reserves is __?__.

If you fill in the blank with a correct answer, then the answer is a solution to the sentence.

Questions covering the reading

1. What is an open sentence?

2-10. Is the given sentence an open sentence?

2. $2 + 3 > 5$ 3. $x + 3 > 5$

4. _____ $\cdot$ $10 = 90$ 5. $9 \cdot 10 = 90$

6. The average person can expect to live at least ? years.

7. _____ causes lung cancer.

8. That tree is y years old.

9. $B = 93.4$ 10. $2A + 14 > 3A$

11. In Questions 9 and 10, no replacement set is given. What replacement is assumed for each variable?

12. Define: solution to an open sentence.

13-18. (a) Name a replacement for the variable which is a solution.
       (b) Name a replacement for the variable which is not a solution.

13. $3 \cdot F < 18$ 14. M has been President of the United States.

15. $D + 4 = 11.3$ 16. $14 > G - 1$

17. $\frac{h}{5} = 10$ 18. x is a prime number.

19-22. Match the sentence with the graph of its solutions. The replacement set for each variable is the set of all real numbers.

19. $-2 \leq w \leq 1$

(a)

20. $-2 < x \leq 1$

(b)

21. $-2 < y < 1$

(c)

22. $-2 \leq z < 1$

(d)

23-28. (a) Assume the set of real numbers is the replacement set for the variable. (b) Assume the set of integers is the replacement set. In each case, graph all solutions.

23. $5 > x$

24. $15 < y$

25. $z \leq -2$

26. $A \geq 3\frac{1}{2}$

27. $2.1 < B < 2.5$

28. $-8 \leq C \leq 10$

Questions testing understanding of the reading

1-4. (a) For each sentence, name three solutions.
(b) Graph all solutions.

1. A speed limit is 55 mph. Then you are speeding if you are driving L miles per hour.

2. A person is a millionnaire if he has over d dollars in assets.

3. x is between 3 and π.

4. y is larger than -3 and smaller than -2.9.

5. I know they have more than 5 and less than 10 children. Given that x is the number of children, graph the solutions to $5 < x < 10$.

6. Given that t is an integer, graph all solutions to $-9 \leq t < 0$.

7-10. Give an example of an open sentence which has:

7. no solution.

8. exactly one solution.

9. exactly two solutions.

10. infinitely many solutions.

102

11-13. Give an example of an open sentence which has exactly one solution that:

11. you know.   12. can only be estimated.

13. it would be nice to know but no one knows it.

14-29. Take a guess on the solution to each sentence.

14. $v - 46.2 = 87.9$ 15. $x + x = 1$

16. At the age of 15, a white boy in the U.S. can expect to live y years.

17. At the age of 15, a white girl in the U.S. can expect to live z years.

18. At the age of 15, a black boy in the U.S. can expect to live b years.

19. At the age of 15, a black girl in the U.S. can expect to live a years.

20. On January 1 of next year, gasoline will cost about c cents a gallon.

21. During the time of Jesus, there were about d people on Earth.

22. $100 - e = 99.9$ 23. $78 + f = 78 + \pi$

24. If you put \$20 in a savings account at 6% yearly interest, at the end of a year you will have _____ in the account.

25. If the floor of a room is rectangular with dimensions 3.2 meters by 4.6 meters, its area is A.

26. There are less than W whooping cranes in the world.

27. S is presently the United States Secretary of State.

28. G is presently the United Nations Secretary-General.

29. X is the most populous country on Earth.

Lesson 8

Ordered Pairs

Many open sentences have more than one blank.

> If one blouse costs $7.99,
>
> then _____ blouses cost _____ dollars.

Two variables are needed.

> If one blouse costs $7.99,
>
> then b blouses cost c dollars.

When b is 5, then c is 39.95. Five blouses cost $39.95 at this price. The two numbers 5 and 39.95 make up <u>one</u> solution to the sentence. This solution is written (5, 39.95), called "<u>the ordered pair</u> 5, 39.95." The order of the variables is usually alphabetical, so b is put first, c second. Here are other solutions to the above sentence.

(10, 79.90) (1, 7.99) (2, 15.98) (3, 23.97) (100, 799.00)

Whereas real numbers are pictured on a number line, ordered pairs of real numbers are pictured in the following manner.

1. Draw a <u>horizontal</u> number line.
    Make a scale on that line.
    This line is called the <u>x-axis.</u>

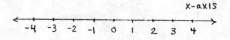

104

2. Draw a <u>vertical</u> number line intersecting the x-axis. This second line is called the <u>y-axis</u>. Make a scale on the y-axis.

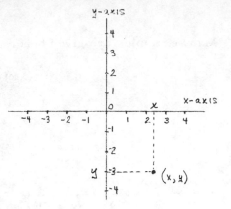

3. To graph (x, y) go to x on the x-axis, y on the y-axis, and lightly draw a rectangle as at right. The 4th vertex is the graph of (x, y).

For obvious reasons   x   is called the <u>x-coordinate</u> of the point (x, y).   y   is the <u>y-coordinate</u> of (x, y).   Together   x   and y   are the <u>rectangular coordinates</u> of (x, y).

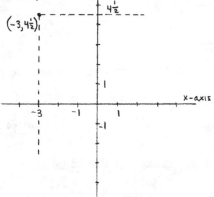

Example: To graph $(-3, 4\frac{1}{2})$:

1. Go to -3 on the x-axis. All points with x-coordinate -3 lie on the dotted line.

2. Go to $4\frac{1}{2}$ on the y-axis. All points with y-coordinate $4\frac{1}{2}$ lie on the dotted line.

3. Graph $(-3, 4\frac{1}{2})$ where the dotted lines intersect.

In mathematical examples, the axes usually intersect at (0,0). This key point is called the <u>origin</u>. The x-coordinate tells how far right or left of the origin the point is graphed. The y-coordinate tells how far up or down from the origin.

Many examples are given here.

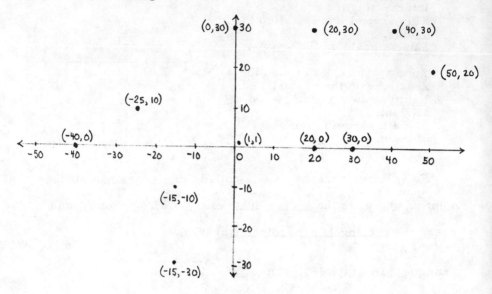

Many relationships can be pictured using ordered pairs. Graphed below is the days overdue - library fines relationship of Question 3, p. 71. Notice the simple pattern of the graph.

Day	Fine	Ordered pair
1	.10	(1, .10)
2	.15	(2, .15)
3	.20	(3, .20)
4	.25	(4, .25)
5	.30	(5, .30)
6	.35	(6, .35)
7	.40	(7, .40)
8	.45	(8, .45)
9	.50	(9, .50)

The exercises show the variety of applications of graphing.

In these applications, you should note:

If you don't have enough room:

1. It is possible (and common) to have different scales on the axes. (The graph on p. 106 is an example.)

2. The 0 point on the x-axis does not have to be the same point as the 0 point on the y-axis.

Questions covering the reading

1-3. Trace the drawing at right. Label the:

1. x-axis.

2. y-axis.

3. point (x, y).

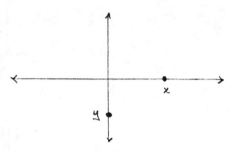

4-13. Use the drawing at left.

Pick the letter which identifies the graph of the given point.

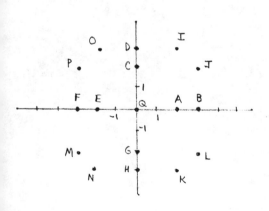

4. (2, 3)   5. (0, 3)

6. (2, 0)   7. (-2, 3)

8. (3, -2)   9. (-3, -2)

10. (0, -3)   11. (0, 0)

12. origin   13. (-2, 0)

14-17. Carefully graph the points on the same pair of axes.

14. (-6, -5), (-3, -5), (0, -5), (3, -5), (6, -5)

15. (2, 11), (2, 6), (2, 0), (2, -1), (2, -4)

16. (-3, 9), (-2, 4), (-1, 1), (0, 0), (3, 9)

17. $(4\frac{1}{2}, 6)$, $(-4\frac{1}{2}, -6)$, $(-4\frac{1}{2}, 6)$, $(4\frac{1}{2}, -6)$

18-21. Draw axes like the ones that are given (but make your drawing larger). Then plot the points on your graph.

18. $(1, \frac{1}{2})$, $(2, \frac{1}{4})$, $(3, \frac{1}{8})$, $(4, \frac{1}{8})$

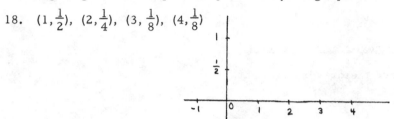

19. (400, -100), (-500, 0), (600, 100), (0, -430)

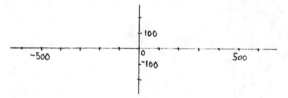

20. (-200, 2), (-300, -3), (-100, 4), (700, 5)

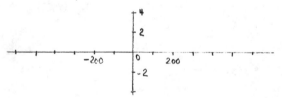

21. (45, 0), (55, 3), (50, -2), (53, 4), (57, -1)

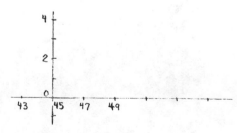

108

22-23. Name 5 solutions to the sentence.

22. If one blouse costs $9.50, then b blouses cost d dollars.

23. There are x centimeters in y meters.

Questions applying the reading

1. Comparing students. Five students take two English tests with the following results.

	Vocabulary	Writing
Pam	85	81
Quincy	78	77
Raoul	80	82
Sara	83	88
Thea	73	90

The results can be written as 5 ordered pairs. Carefully graph these pairs using axes labelled like those above. From your graph, determine which students would probably be put in the same class and which two students would most likely be in different classes.

2. Locations on Earth. Compare the locations of these U.S. cities by graphing. (When cities are close, this idea gives an accurate picture.) Numbers given are approximations. Consider west as negative. (Otherwise your drawing will be backward.) A hint for the graphing is given.

	W. Longitude	N. Latitude
New York	74	41
Washington	77	39
Chicago	88	42
Houston	95	30
Denver	105	40
Los Angeles	118	34

3. <u>Caribbean Resort Islands.</u> Follow the directions of Question 2 to find out where these islands are located.

	W. Longitude	N. Latitude
Nassau, Bahamas	77	25
Bermuda	65	32
Kingston, Jamaica	77	18
San Juan, Puerto Rico	66	18.5
Saint Croix	65	18
Martinique	61	15

(For reference, Miami is near 80°W, 26°N.)

4. Life Expectancy. (a) Accurately graph the data given here. (Source: U.S. Dept. of Health, Education and Welfare, National Center for Health Statistics, 1971 Data) Use axes like those labelled. (b) The life expectancy is the average number of years you could expect to live if born at that time. What do you think the female life expectancy would be in 1980? Graph the point corresponding to your guess.

No. of years ago (from 1975)	U.S. Female Life Expectancy
75	48.3
65	51.8
55	54.6
45	61.6
35	65.2
25	71.1
15	73.1
5	74.6

5.

No. of years ago (from 1975)	U.S. Male Life Expectancy
75	46.3
65	48.4
55	53.6
45	58.1
35	60.8
25	65.6
15	66.6
5	67.1

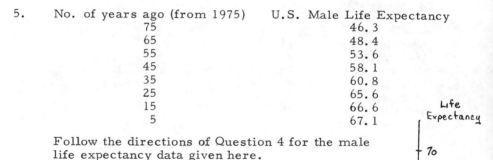

Follow the directions of Question 4 for the male life expectancy data given here.

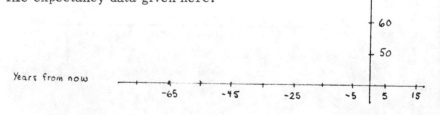

110

6. <u>Geometric Figures</u>. The orbit of the Earth around the sun is an ellipse. Ellipses are hard to draw. Here are 16 points which lie on an ellipse. Accurately graph the points. Then connect them in order by a smooth curve. Use axes with the same scale which intersect at (0, 0).

(-10, 1) (7, 8) (10, 9) (23, 12) (25, 12) (26, 9) (25, 8) (14, 1) (10, -1)

(-7, -8), (-10, -9) (-23, -12) (-25, -12) (-26, -9), (-25, -8) (-14, -1)

7. Follow the directions of Question 6. What figure can be formed?

(3, 4) (4, 3) (5, 0) (4, -3) (3, -4) (0, -5) (-3, -4) (-4, -3) (-5, 0)

(-4, 3) (-3, 4) (0, 5)

8. <u>Describing Objects</u>. When pictures are not available, coordinates can sometimes describe a figure more quickly than other methods. Suppose over the phone you wish to describe the miniature golf hole at right. This can be done by giving coordinates of the 7 points. Let point A be (0, 0) and suppose that B is on the y-axis. Give the coordinates of the other points.

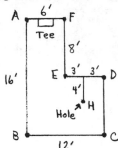

9. Follow the directions of Exercise 8 for this piece of wood used in building. Dimensions are in meters. Pick a convenient origin.

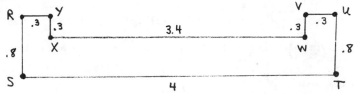

10-11. <u>Solutions</u>. Graph 10 solutions to the two-variable sentences.

10. If one record costs $4.50, then n records cost c dollars.

11. If the area of a rectangle is 24 square centimeters, then the width could be w and the length ℓ .

12. <u>Frequencies</u>. A person tosses a die 120 times, with the following results:

Lands on:	1	2	3	4	5	6
No. of times:	18	23	21	15	24	19

Graph the six ordered pairs suggested by this data.

13. Graph the relative frequencies suggested by Question 12.

<u>Questions for discussion</u>

1. Rectangular coordinates are sometimes called <u>Cartesian</u> coordinates, named after the French mathematician and philosopher René Descartes who first used them. Descartes (1596-1650) used the Latin name Cartesius. Why would his Latin name be so important?

2. Places on maps are often located by letters and numbers written at the side of the page. Why is this method not as good as the use of rectangular coordinates?

Lesson 9

<u>Subscripts</u>

A subscript is a symbol which is written to the right and below a second symbol. The symbol 3, A, and 0 below are subscripts.

$$x_3 \qquad b_A \qquad z_0$$

Read: x sub three   b sub A   z sub 0
             or
           x three

112

<u>First use of subscripts</u>: Abbreviations

In Lesson 6, the symbols L_M and L_I stood for "length in meters" and "length in inches." Here the subscripts M and I are abbreviations for meters and inches.

<u>Second use of subscripts</u>: Helping to name variables

Suppose a stereo dealer has 5 stores. The stores could be numbered from 1 to 5. Then the sales and profits (in dollars) from the stores could be identified as follows:

S_1 = sales from 1st store P_1 = profit from 1st store

S_2 = sales from 2nd store P_2 = profit from 2nd store

S_3 = sales from 3rd store P_3 = profit from 3rd store

S_4 = sales from 4th store P_4 = profit from 4th store

S_5 = sales from 5th store P_5 = profit from 5th store

For example, if the 3rd store had $5000 worth of sales and a loss of $300, then $S_3 = 5000$ and $P_3 = -300$.

Here the letters S and P by themselves are not variables. The ten variables are S_1, S_2, S_3, S_4, S_5, P_1, P_2, P_3, P_4, P_5. The subscripts help to name the variables. There was a simple pattern used in naming these variables. The subscript n stands for the number of the store.

S_n = sales from nth store P_n = profit from nth store

(The word <u>nth</u> is pronounced "enth.")

Without further information there is no way of knowing how or if the variables S_1, P_1, S_2, P_2, ... are related.

<u>Third use of subscripts</u>: Naming lots of variables

Some situations require hundreds of variables. (See Questions 7-10 below.) You would run out of letters if subscripts were not used.

<u>Fourth use of subscripts</u>: Ordering variables

In graphing, (x, y) often stands for a point. If there are two points, they can be called (x_1, y_1) and (x_2, y_2). In this situation x_1 means "x-coordinate of 1st point." x_2 means "x-coordinate of 2nd point.

A third point could be called (x_3, y_3). Without further information there is no way of know if x_1, x_2, x_3, y_1, y_2, or y_3 are related.

<u>Questions covering the reading</u>

1.   Name the subscript in A_b.

2.   Which is <u>not</u> true? Subscripts are used:

   (a) to help name variables.   (b) for ordering variables.
   (c) for abbreviation.   (d) for underwater writing.

3-6.   Pronounce each out loud.

3. Z_0 4. S_3 5. t_n 6. nth

7-10.   Five hundred students, numbered 1 to 500, take a test. The score of the nth student is called S_n.

7.   What does S_5 stand for?

8.   What is the score of the 133rd student?

9.   True or False: $\dfrac{S_3 + S_5}{2} = S_4$.

10.   If it is possible, calculate S_{27}.

11-16. The stereo dealer mentioned on p. 113 had the following approximate sales and profits one day from his 5 stores.

Store	Sales	Profit
1	8250	1000
2	11000	2500
3	5000	-300
4	8800	800
5	3290	-1100

From this information, calculate:

11. S_2 12. P_5 13. P_1 14. S_4

15. $S_1 + S_2 + S_3 + S_4 + S_5$ 16. $S_4 - P_4$

17-20. Use the information of Questions 11-16.

17. What does the answer to Question 15 stand for?

18. What do the negative values of P_3 and P_5 mean?

19. Is it true that $2S_2 = S_4$?

20. Is it true that $P_1 + P_2 = P_3$?

21-28. Here are 4 points in order: (1, 2), (3, 5), (-7, -2), (60, -4). Let x_n be the x-coordinate of the nth point. Let y_n be the y-coordinate of the nth point. Calculate:

21. x_1 22. x_3 23. y_2 24. y_4

25. $3x_4$ 26. $y_2 - y_1$ 27. $x_4 - 1$ 28. $x_1 + 3x_2$

1-8. Suppose you are doing an experiment with 4 mice, numbered 1 to 4.

	Beginning weight	Final weight
Mouse No. 1	200 g	320 g
Mouse No. 2	175 g	150 g
Mouse No. 3	180 g	250 g
Mouse No. 4	190 g	140 g

Let B_n and F_n stand for the beginning and final weights (in grams) of Mouse n. Then what is:

1. B_1 2. F_1 3. $4B_3$ 4. $7F_2$

5. B_5 6. $F_1 + F_3$ 7. $F_3 - B_3$ 8. $\frac{1}{2}(B_2 + B_4)$

9-10. You put \$100 into a savings account. Suppose you withdraw \$5 every week. Then let L_n be the amount left in the account after n weeks. (We ignore interest here.)

9.   Calculate L_1, L_2, L_3, L_4, and L_5.

10.   Multiple Choice. What is a formula for L_n?

   (a) $L_n = 100n$ (b) $L_n = 100 - 5n$

   (c) $L_n = 5n + 95$ (d) $L_n = 90 + 10n$

11-13. Twelve students measured their feet (as in Question 5, p. 55) and got the following measurements:

   11", 13", 8", 9.75", 13.5", 9.75", 8", 9", 12", 9", 8", 10.5"

11.   Let S_n be the length of the foot of the nth student. What is S_{10}?

C 12.   Find the mean of all the S_n.

13.   Do you think there is a simple formula for S_n?

14-26. Here are scores of 6 students on two tests

Student	Before studying	After studying
1	60	65
2	50	85
3	80	85
4	73	91
5	57	75
6	89	97

Let B_n be the score of the nth student before studying.

Let A_n be the score of the nth student after studying.

14. What is B_3?   15. What is A_5?

16. Calculate $A_2 - B_2$.   17. Calculate $A_6 - B_6$.

18. In general, what does $A_n - B_n$ stand for?

19. Calculate $\dfrac{A_1 + A_2 + A_3 + A_4 + A_5 + A_6}{6}$ and tell what your answer

   means.

20. Repeat Question 19 for $\dfrac{B_1 + B_2 + B_3 + B_4 + B_5 + B_6}{6}$.

21-23. (The letter i is often used as a subscript in place of n.)
Suppose S_i is the score of the ith student before studying. Calculate:

21. $5S_4$ 22. $2S_6 + S_3$ 23. $S_2 + 3$

24-26. (The letter k is often used as a subscript in place of n.)
Suppose T_k is the score of the kth student after studying. Calculate:

24. $3T_2 + 3T_1$ 25. $100 - T_6$ 26. $200 - 2T_5$

Mathematics is the study of patterns. In order to describe these patterns, <u>variables</u> are often needed. A <u>formula</u> is a description of a pattern which displays one variable in terms of other variables.

Letters of the alphabet are usually used to name variables. When two variables are being discussed, their values are often placed in <u>ordered pairs</u>. When variables need to be ordered or when many variables need to be described, <u>subscripts</u> are helpful. The number 3 is the subscript in the variable A_3.

To avoid confusion when mathematical symbols (including variables) are used, there are rules for the language of mathematics which are followed world-wide: work in parentheses first, then multiplications or divisions from left to right, then additions or subtractions from left to right.

A <u>variable</u> is a symbol which can stand for any one of the elements of a given set, called the <u>replacement set</u> of the variable. The replacement or "thing stood for" is called a <u>value</u> of the variable.

An _open sentence_ is a sentence which has variables in it. A _solution_ to an open sentence is a value of the variable which makes the sentence true.

In short, this chapter has discussed three uses of variables:

> to describe patterns
> in formulas
> in open sentences

Graphing can lead to geometrical descriptions of many patterns, formulas, and solutions to open sentences, so is related to each of these uses of variables.

CHAPTER 3

ADDITION AND SUBTRACTION

Lesson 1

Models for Addition

The operation called <u>addition</u> is important because adding gives answers in many actual situations. It is impossible to list all of these situations. So we categorize them. The easiest way to categorize the situations is by the uses of the numbers being added. Numbers in addition problems are almost always used for

(1) counting

or (2) measuring

or (3) situations with two opposite directions

A <u>model</u> for an operation (like addition) is a general pattern which includes many of the uses of the operation. Because numbers in adding are used in three different ways, there are three models for addition.

Addition Model 1 (from counting): Union

You probably first learned to add by counting. A typical situation is as follows: Suppose you have 8 pieces of candy and a friend has 5. How many pieces are there altogether?

Counting can give the answer. There are 13 pieces altogether. Addition is a shortcut. $8 + 5 = 13$.

The union of two sets S and T, written S∪T, is the set of all elements in either S or T. For example,

if $S = \{1, 2, -3, 8, 21\}$, $T = \{4, 8, 12, 2\}$
then $S ∪ T = \{1, 2, -3, 8, 21, 4, 12\}$

If set A has 8 elements and set B has 5 elements and there is no overlap, then A ∪ B will have 13 elements. This is the same addition problem as with the candy.

Variables are needed to describe the general pattern. The language of sets helps make the description precise.

<table>
<tr><td>Union Model
for Addition:</td><td>If set A has a elements and set B has b elements, and A and B have no elements in common, then the union A ∪ B has <u>a + b</u> elements.</td></tr>
</table>

The union model for addition only fits adding natural numbers and 0. No set has π elements. You cannot count 1.3 elements. A more general model applies to all nonnegative numbers. This model involves things which are measured.

Addition Model 2 (from measuring): Joining

Segment $\overline{PQ}$ has length 2. 6 cm. Segment $\overline{QR}$ has length 5. 3 cm. If Q is between P and R on the same line, what is the length of PR?

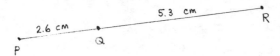

Drawing is not actual size.

Counting does not work for these lengths. But you can still <u>add</u> to get the length of $\overline{PR}$.   2. 6 + 5. 3 = 7. 9, so 7. 9 is the length of PR.

Of course if B is not between points A and C, adding will <u>not</u> give you the length of $\overline{AC}$.

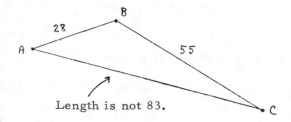

Length is not 83.

When you buy groceries, you know that the prices are added to get the total price you pay. This is because each price can be thought of as the length of a segment.

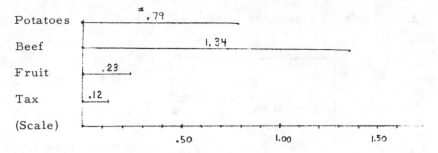

To total up the prices, lay the segments end to end.   The total length
is found by adding.

.79   1.34   .23   .12

Total length = .79 + 1.34 + .23 + .12 = 2.48

You can measure length.   You can also measure weight, area,
force, volume, etc.   In all these cases:

Joining Model
for Addition:

> If something with measure   a   is joined
> to something with measure   b,   the result
> has measure a + b.

Examples: Joining Model of Addition

1.   If you weigh 125 lb. and a friend weighs   x   lb. and you step on a
scale together, the total weight is 125 + x lb.

2.   The total area below is 50.6 + 8.3 or 58.9 sq cm.

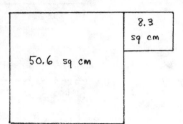

Drawing is not actual size.

Measuring can involve any positive number or 0. So the joining model does not show how to add negative numbers. A still more general model - the slide model - is needed. This model for addition is discussed in Lesson 2.

Models are important for three reasons. First, a model can be used to check your answer to a problem. (For example, you can check the answer to 8 + 5 by actually counting. You can check an answer to 3.24 + 8.91 by drawing line segments and measuring.) Second, the models for an operation tell you when to use that operation. (For instance, if you have a problem and are not sure whether to add or multiply, a model may help you make that decision.) Third, a model for an operation can show properties of the operation. (For example, segments of lengths a and b can be put together in either order giving the same total length. From this measure model, we know that a + b = b + a. That is, addition of nonnegative numbers is commutative.) For these reasons, it is very important that you learn the models for operations.

124

1. What uses of numbers are found in adding situations?

2. Define: model for an operation.

3. Give an example of a counting problem which can be done by addition.

4. Describe the union model for addition using variables.

5-7. Let A = $\{1, 2, 3, 4, 5\}$, B = $\{4, 6, 8\}$, C = $\{0, 1, 2\}$. Write down:

5. A $\cup$ B   6. A $\cup$ C   7. B $\cup$ C

8. Which two sets of Questions 5-7 can be used to give an example of the union model for addition? Why can they be used?

9. Describe the joining model for addition. (Use variables.)

10-12. Give an example of the joining model for addition where the things being measured are:

10. weights   11. distances   12. areas

13-16. Points A, B, C, and D lie on the same line. Give the length of:

13. $\overline{BD}$

14. $\overline{AC}$

15. $\overline{AD}$

16. $\overline{ED}$

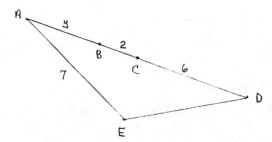

17-19. There are three reasons why models are important. They are named here. Give an example for each reason.

17.   Models help check answers.

18.   Models can tell you when to use an operation.

19.   Models can show properties of an operation.

Questions testing understanding of the reading

1.   Trace these segments. (a) Draw a segment of length c + d.
      (b) Draw a segment of length d + c.

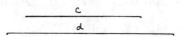

2.   (You will need a ruler.) Given is a segment with length x cm.
      Draw a segment of length x + 3 cm.

3.   Marty gets on a scale and weighs x kg. Jerry gets on a scale and weighs 60 kg. Together they weigh 112 kg. How are x, 60 and 112 related?

4.   The total area of the apartment is given as 70 square meters. Figure out something about A_1 and A_2.

      Use the figure at right.

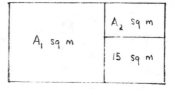

5.   Show that shopping bills are an application of the joining model for addition by drawing segments of the given lengths and then estimating the bill by finding the total length.

      Eggs $.89   Meat $2.17   Vegetables $.75   Tax $.19

6-8. Which model of addition is used in each situation?

6.   5 teaspoons of salt are added to 2 1/2 teaspoons of salt, resulting in 7 1/2 teaspoons of salt.

126

7.   The Great Lakes have the following areas:

Superior	82,400 sq km
Huron	59,700
Michigan	58,000
Erie	25,700
Ontario	19,700
Total area	245,500 sq km

8.   From 1930 to 1967, there were the following numbers of executions in the U.S.

White	1751
Negro	2066
American Indians or Orientals	42
Total	3859

(There have been no executions from 1967 until the time of this writing, June, 1976.)

Questions for discussion

1.   When 1/3 cup of sugar is added to 2/3 cup of coffee, the result does not fill up a cup.   Why not?

2.   Suppose a quart of oil and a quart of water are mixed.   Do two quarts of liquid arise?

Lesson 2

The Slide Model for Addition

 Numbers being added often represent situations with two opposite directions. We may add gains and losses in a football game, deposits and withdrawals in savings accounts, ups and downs of temperatures, profits and losses in business, and so on. Each of these situations uses both positive and negative numbers.

Addition Model 3 (from situations with opposite directions): Slides

 Think of positive numbers as being slides in one direction. Think of negative numbers as being slides in the opposite directions. Here are slides for 5 and $-3\frac{1}{2}$. Arrows show the direction.

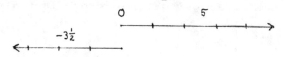

Now start the second slide where the first is left off. This is following one slide by the other. If a slide 5 in one direction is followed by a slide $3\frac{1}{2}$ in the opposite direction, you wind up $1\frac{1}{2}$ in the first direction. The answer is dotted.

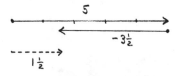

In the slide model, the "+" of addition means "followed by" in the real world. The arrows show the addition $5 + -3\frac{1}{2} = 1\frac{1}{2}$.

Examples: Adding a positive and a negative number.

1. Banking: A deposit of $5 followed by a withdrawal of $3.50 leaves $1.50.

$$5 \quad + \quad -3\tfrac{1}{2} \quad = \quad 1\tfrac{1}{2}$$

2. Football: A five-yard gain followed by a $3\tfrac{1}{2}$ yard loss results in a $1\tfrac{1}{2}$ yard gain.

3. Temperature: If a temperature of -10° goes up 25°, the new temperature is 15°.

$$-10 \quad + \quad 25 \quad = \quad 15.$$

4. Suppose you buy 25 bottles of soft drinks for a party. You use 10. Then you will have 15 extra.

If both slides are in the same direction, the result is a slide

in that direction.

$$-10 + -5 = -15$$

$$21.2 + 10.5 = 31.7$$

Examples: Adding negative numbers

1. If you owe $250 to one store and $120 to another, you owe $370 altogether.

$$-250 \quad + \quad -120 \quad = \quad -370$$

2. Let travelling east be considered positive. Suppose you travel 250 miles west. Then you travel 120 miles west. You will have travelled a total of 370 miles west.

Of the three models for addition, the slide model is the most general. So it is often the most useful.

| Slide Model for Addition: | If a slide of a is followed by a slide of b, then the result is a slide of a + b. |

Questions covering the reading

1-3. Addition has three models:

                Union   Joining   Slide

1.   Which model applies to all real numbers?

2.   Which model can only be applied to nonnegative integers?

3.   Which model can only be applied to positive real numbers?

4.   What is the slide model of addition?

5-8. Here is an arrow to represent the number 10.

10

Use this scale and draw an arrow to represent each number.

5.   20   6.   -20   7.   -5   8.   30

9-12. Draw arrows to represent each addition. Use the arrow of Questions 5-8 as a guide.

9.   10 + 10 = 20   10.   20 + -20 = 0

11.   10 + -25 = -15   12.   -5 + -10 = -15

13-15. Show how the addition in Question 11 could come from each type of situation.

13.   football   14.   banking   15. temperature

16-25. Give the result of each. Then write the problem and answer using only mathematical symbols.

16. a gain of 6 yds. followed by a loss of 2 yds

17. a loss of $11.34 followed by a loss of $6.93

18. going up 270 meters on a mountain one day, making no gain the next

19. winning $600, then losing $2

20. losing $600, then winning $2

21. going east 43 miles followed by going west 100 miles

22. withdrawing $49.26, then depositing $12.47

23. going north 12.5 km, then going south 3.25 km

24. withdrawing $11, then withdrawing $7.12

25. a temperature going up 6°, then going down 6°

Questions testing understanding of the reading

1-4. (a) Translate into a real-life situation. (b) Perform the addition.

1. $^{-}3 + 11$ 2. $^{-}14.22 + {}^{-}7.30$

3. $^{-}246 + {}^{-}2$ 4. $^{-}200 + {}^{-}200$

5-17. Simplify.

5. $^{-}10 + 520$ 6. $520 + {}^{-}10$ 7. $6 + {}^{-}20$

8. $^{-}16 + {}^{-}5$ 9. $^{-}50 + 37$ 10. $^{-}1.4 + 3.0$

11. $\frac{1}{2} + {}^{-}\frac{3}{2}$ 12. $^{-}\frac{3}{4} + {}^{-}\frac{5}{4}$ 13. $^{-}\frac{5}{2} + \frac{3}{4}$

14. $^{-}46 + {}^{-}32 + 32$ 15. $87.2 + {}^{-}.9 + 15$

16. $^{-}.9 + 87.2 + 15$ 17. $\frac{1}{2} + 0 + {}^{-}\frac{3}{2}$

18-23. These addition problems are more difficult.

18. .346 + -1   19. 3.241 + -10   20. -.34 + .012

21. $-\frac{1}{2} + \frac{3}{5}$ 22. $\frac{20}{3} + -\frac{7}{4}$ 23. $-\frac{11}{6} + -\frac{1}{6}$

24. Show using slides: a loss of $500 followed by a gain of $600 followed by a gain of $350 followed by a loss of $200. What is the total of all these?

25. Graph these 5 points: (-2, 5) (1, 6) (2, 5) (3, -1) (0, 0). Connect the points to form a 5-sided polygon (pentagon). Now add -3 to the first coordinate of each point and graph the 5 new points. What has happened?

Lesson 3

The Assemblage Property of Addition

A person makes deposits and withdrawals from a savings account.

		Deposit	Withdrawal
Sept.	1	$150.00	
	10		$85.00
	15	$200.00	
	20		$100.00
	21		$25.00
Oct.	1	$175.00	

At the end of the day on October 1st, how much is in the account? This is a problem in addition. Six numbers are to be added. Three are

132

positive (deposits).   Three are negative (withdrawals).

The bank adds in chronological order:

$$150 + {}^-85 + 200 + {}^-100 + {}^-25 + 175$$

$$= \quad 65 + 200 + {}^-100 + {}^-25 + 175$$

$$= \quad 265 + {}^-100 + {}^-25 + 175$$

$$= \quad 165 + {}^-25 + 175$$

$$= \quad 140 + 175$$

$$= \quad 315$$

It is easier to add all deposits, then all withdrawals:

$$(150 + 200 + 175) + ({}^-85 + {}^-100 + {}^-25)$$

$$= \quad 525 \quad + \quad {}^-210$$

$$= \quad 315$$

Changing the order of the numbers and addtitions does not affect the result.   At the end of October 1st, \$315 is in the account. This is an instance of the following pattern:

> Given numbers to be added,
> changing the order of the
> numbers or the order of addi-
> tions does not affect the sum.

In this book, this property is called the <u>assemblage property</u> of addition. The assemblage property has two parts to it:

Order of terms can be switched:
$$413 + 229 = 229 + 413$$

We say that addition is <u>commutative.</u> (The French word "commutatif" means switchable. It was first used by Francois Servois in 1814.)

Order of additions makes no difference: Given $3 + 2 + 17$

left addition first	right addition first
$(3 + 2) + 17$	$3 + (2 + 17)$
$=\ \ 5 + 17$	$=\ \ 3 + 19$
$=\ \ 22$	$=\ \ 22$

We say that addition is <u>associative</u>. (The word was first used by the English mathematician William Rowan Hamilton in 1835.)

When an operation is commutative and associative, it has the assemblage property. The numbers may be assembled in any way before applying the operation. Multiplication also has this property.

But not all operations have the assemblage property. In subtraction, order of terms makes a difference.

$$3 - 2 \text{ is } \underline{\text{not}} \text{ equal to } 2 - 3$$

Order of subtractions also makes a difference.

$$(3 - 2) - 1 \text{ is } \underline{not} \text{ equal to } 3 - (2 - 1)$$

So in doing subtraction, you must be careful.

Questions covering the reading

1-4.   Give an example to back up each statement.

1.   Addition is commutative.   2.   Subtraction is not commutative.

3.   Addition is associative.   4.   Subtraction is not associative.

5.   The French word "commutatif" means _____.

6.   In what century were the "commutative" and "associative" properties named?

7.   If an operation is associative and commutative, then it has the _____ property.

8.   What is the assemblage property of addition?

9.   Give an example of the assemblage property of addition.

10.   Name an operation which does not have the assemblage property.

1.   A person has a checking account. On October 31st, there was
      $312.47 in the account. Here are deposits and checks written
      during the first two weeks of November. How much was in the
      account as of November 16th?

			Checks	Deposits
Nov.	1	Paycheck		$325.00
	2	Gas bill	$37.22	
	4	Electric bill	$38.18	
	4	Cash	$100.00	
	6	Cash	$75.00	
	9	Phone bill	$31.22	
	11	Gift		$50.00
	12	Cash	$100.00	
	15	Paycheck		$325.00

2.   Between 1960 and 1970, Blacks moved in and out of the South
      Atlantic States in the following numbers:

West Virginia	-20000	
Delaware	4000	
Maryland	79000	Negative numbers
Georgia	-154000	mean moving out.
District of Columbia	36000	
Virginia	-79000	Positive numbers
North Carolina	-175000	mean moving in.
South Carolina	-197000	
Florida	-32000	

      What was the total migration of Blacks in these states in those
      years?

3.   A professional golfer plays 6 rounds of golf with the following
      results: 4 under par, 3 over par, par, 2 over par, 2 under par,
      1 under par. How did she finish with respect to par?

136

4. In bowling, total pins scored often wins a match. Suppose a match is 6 games. If a bowler loses the first game by 23 pins, wins the second by 17, wins the third by 7, loses the fourth by 13, loses the fifth by 16, and wins the sixth by 11, by how much has he won or lost the match?

5. A couple made a budget. Each month they figured out whether they spent more or less than their budget would allow.

Jan.	$12.50 under	Apr.	$8.00 under
Feb.	$30.00 over	May	$11.40 over
Mar.	$22.60 under	June	$3.20 over

After 6 months, how were they doing?

6-9. Add the following numbers.

6. -11.9, -13, 6.1, -3.2, 8.74, -6.09

7. $\dfrac{2}{5}$, $-\dfrac{3}{5}$, $\dfrac{11}{5}$, $-\dfrac{12}{5}$, $\dfrac{6}{5}$, $-\dfrac{7}{5}$

8. 3, -3, -4, 6, 11, -11, -6, 0

9. 6132, -8394, -142, -9038

10-12. What property of addition is being used?

10. Check addition of a column of numbers by adding up instead of down.

11. To add 3 + -4 + 4 you add -4 + 4 first.

12. Realize that $469\frac{1}{2}$ is $\frac{1}{2}$ + 469.

13-18. (a) Is the commutative property being applied?
(b) Is the associative property being applied?
(c) Is the assemblage property being applied?

13. 3x + 4 = 4 + 3x

14. (2+9) + (5+6) = (5+6) + (2+9)

15. $\frac{1}{2} + (\frac{1}{2} + \frac{1}{3}) = (\frac{1}{2} + \frac{1}{2}) + \frac{1}{3}$

16. a + b + c = b + c + a

17. -2 + -8 + 5 + 6 = -2 + 6 + 5 + -8

18. x + y + z + 5 = x + (y + z) + 5

Question for discussion

1. How is "commutative" related to the word "commuter"?

137

Lesson 4

Zero and Addition

If you make no withdrawal or deposit, the amount in an account stays the same.

$$\text{amount} + 0 = \text{amount}$$

It is like "no gain" in football. You stay where you are. Adding 0 to a number keeps the identity of that number. So the number 0 is called the <u>additive identity</u>.

A deposit of \$50 followed by a withdrawal of \$50 leaves you where you were at the beginning. In symbols,

$$50 + {}^-50 = 0$$

Walking x km east, then x km west ($^-$x km east) gets you back where you started. x + $^-$x = 0. This is the property of opposites.

Here are the four properties of addition studied in this and the last lesson. All are obvious from the slide model.

```
Properties of Addition of Real Numbers


For any real numbers  a,  b,  and  c:


(Assemblage Properties)
1.  commutative property            a + b = b + a
2.  associative property         (a + b) + c = a + (b + c)

(Properties of Zero)
3.  additive identity property        a + 0 = a
4.  property of opposites            a + -a = 0
```

Questions covering the reading

1-11. Which property of addition is described?

1. Order of additions can be changed without affecting the sum.

2. If a number is added to 0, the sum is that number.

3. Gain 10 yards, then lose 10 yards, and you will be back where you started.

4. Deposit $50, make no withdrawal, and you have deposited $50.

5. Order of numbers may be changed without affecting the sum.

6. $-27 + -(-27) = 0$

7. $0 = 3m + -3m$

8. $(a + c) + b = (c + a) + b$

9. $-46293 + 0 = -46293$

10. $2x + 0 = 2x$

11. $\frac{1}{2} + -\frac{2}{3} = -\frac{2}{3} + \frac{1}{2}$

1-4.   Evaluate each expression when x = 3 and y = -4.

1.   x + y + -x + -y

2.   (5x + y) + (y + 5x)

3.   -3 + x + -4 + -y

4.   -(-x) + -x + y

5.   What property is used in each step?

(a) 5 + (-5 + x) = (5 + -5) + x
(b)   = 0 + x
(c)   = x

6.   What is the opposite of 0?

Lesson 5

Subtraction

You probably first learned how to subtract with questions like this one.

Question: If you have 7 pennies and take
                 2 away, how many are left?

Answer: 7 - 2 or 5.

In Lesson 2, the problem would have been done by adding 7 and -2, again giving 5. Many real problems can either be done by addition or subtraction.

140

Situation	By Subtraction	By Addition

1. A temperature of $-5°$ falls $3°$.
   What is the new temperature? $-5 - 3$ $-5 + -3$

2. You have $29.50 and spend $2.36
   How much is left? $29.50 - 2.36$ $29.50 + -2.36$

3. The general pattern
   You have a and lose b.
   How much is left? $a - b$ $a + -b$

Remember that subtraction is not commutative or associative and does not possess the assemblage property. The above general pattern suggests that every subtraction problem can be converted to an addition problem. This is particularly helpful because:

Addition is easy.

Addition has many nice properties.

So we define how to do subtraction by converting all subtractions to additions.

Definition of
Subtraction:

$$a - b = a + -b$$

In words, subtracting b is the same as adding the opposite of b.

Examples: Subtraction.

1. $-3 - 10 = -3 + -10 = -13$

   Think: Subtracting 10 is like adding -10.

2. $-47 - -3 = -47 + 3 = -44$

   Think: Subtracting -3 is like adding 3.

3. $1.1 - .8 - 1.4 + 1.2 - 3.1 + .6 - .2$

$= 1.1 + -.8 + -1.4 + 1.2 + -3.1 + .6 + -.2$ (Applying the definition of subtraction)

$= (1.1 + 1.2 + .6) + (-.8 + -1.4 + -3.1 + -.2)$ (Applying the assemblage property)

$= \quad\quad 2.9 \quad\quad + \quad\quad -5.5$

$= -2.6$

4. $-103 - -103 = -103 + 103 = 0$

In general, any number minus itself is 0. Why? Because
$a - a$ is equal to $a + -a$, which (applying the property of
opposites) is 0.

5. $11 - 15 = 11 + -15 = -4$

Notice that $15 - 11 = 4$ and $11 - 15 = -4$. So $15 - 11$ and $11 - 15$
are opposites. This pattern holds for any real numbers:
$a - b$ and $b - a$ are opposites. Why? Because they add to 0.
That is,

$$(a - b) + (b - a) = a + -b + b + -a$$
$$= a + -a + -b + b$$
$$= \quad 0 \quad + \quad 0$$
$$= \quad\quad 0$$

You now have seen three uses of the - sign.

    $-x$ in front of a variable, say "opposite of"

    -3 in front of a positive number, say "negative"
or "opposite of"

    $2 - 5$ between two numbers or variables, say "minus"

Only the last use of - refers to subtraction.

Subtraction has important applications. Some are given in Lesson 6.

142

1-4. Give (a) a subtraction problem and (b) an addition problem which will answer the given question. Then (c) answer the question.

1. A person has $25.00 in his checking account and mistakenly writes a check for $40.00. By how much is the account overdrawn?

2. A football team, having a bad day, has $^-5$ in net yardage. If the team loses 10 yards on the next play, what is its new net yardage?

3. The temperature of a fluid is $^-173^\circ$ and goes down 1°. What is the new temperature?

4. Lauren earns $25.00 and spends $20.37. How much is left?

5. Subtracting 6 gives the same result as adding _____.

6. Subtracting $^-71.3$ gives the same result as _____.

7-9. Translate into English. No mathematical symbols allowed.

7. $^-34$ 8. $^-2 - 5$ 9. ^-y

10. Pick all correct answers: In general, $x - y =$

   (a) $x + {}^-y$ (b) $^-y + x$ (c) $x + y$ (d) $y - x$

11. Pick all correct answers: $a - b$ and $b - a$:

   (a) unequal   (b) are opposites   (c) add to zero

12. If $a - b = 40$, then $b - a =$ _____.

13-28. Simplify.

13. $17 - 17$ 14. $17 - 18$ 15. $17 - 19$ 16. $^-3 - {}^-3$

17. $^-4 + {}^-4$ 18. $^-4 - {}^-4$ 19. $\dfrac{1}{2} - \dfrac{5}{2}$ 20. $1 - 2$

21. $1.6 - 1.64$ 22. $0 - 8$ 23. $^-2 - {}^-3$ 24. $^-5 - 15$

25. $.3 - 1$ 26. $68 - 92$ 27. $^-7 - {}^-50$ 28. $6 - {}^-6$

29. Why is it helpful to convert subtractions to additions?

30. True or False? When x stands for a negative number, $x - x = 0$.

Questions testing understanding of the reading

1. Seeing the problem   832
                   -469

   Anne says, "That's an addition problem." Betty says, "That's a subtraction problem." Who is right?

2. Suppose you know the answer to 832 - 469. How can you use the answer to calculate 469 - 832?

3-8. Calculate.

3.   2 - 3 - 10   4.   9 - 6 + 11   5.   0 - 9 - 9

6.   -11 + 2 - 5 - 9   7.   10 - (3 - 4)   8.   -10 -8 + 2

9-20. Let $a = 4$, $b = -5$, $c = 6$, and $d = -7$. Calculate.

9. $a - b$ 10. $c - a - b$ 11. $-b - 3 + 2c$

12. $b - 5c$ 13. $c - -d$ 14. $-b - -a$

15. $7 - 2 (b - b)$ 16. $d - 2c + 5a + b$ 17. $b - (a - c) - d$

18. $-c - d - a - b - 11$ 19. $b + b - b - d + d$

20. $(a - b) + (b - a) + 3c - 4a$

21-22. Fill in the table.

			Question 21		Question 22	
If	a =	b =	then a - b =	b - a =	2a - b =	b - 2a =
	5	3				
	5	2				
	5	1				
	5	0				
	5	-1				
	5	-2				

23. Generalize the results of Question 21.

24. Multiple Choice. If 14 - x = 16, then x =

    (a)   30   (b)   -30   (c)   2   (d)   -2

25. <u>Multiple Choice</u>. If $^-100 = y - 600$, then y =

    (a) 30   (b) 500   (c) $^-500$ (d) 700

26. The formula P = s $^-$ c, Profit = selling price $^-$ cost, was used in Chapter 2. If the selling price is $20.95 and the cost to the dealer is $23.17, calculate P. What does your answer mean?

27. Repeat Question 26 if the selling price is $400 and the dealer's cost is $420.

28-30. Three out of three correct is very good on these. Calculate.

28. $^-1.98 - .34$ 29. $^-6 - ^-.01$ 30. $\dfrac{1}{2} - \dfrac{11}{3}$

31. Give the property or definition applied in each step. This shows that x $^-$ y and y $^-$ x are opposites.
    (a) $(x - y) + (y - x) = (x + {}^-y) + (y + {}^-x)$
    (b) $\qquad\qquad\qquad = (x + {}^-x) + (y + {}^-y)$
    (c) $\qquad\qquad\qquad = \quad 0 \quad + \quad 0$
    (d) $\qquad\qquad\qquad = \qquad 0$

Lesson 6

Models for Subtraction

There is a model for subtraction corresponding to each model for addition.

Model for addition	Model for subtraction
1. union	take-away
2. joining	cutting off
3. slide	directed distance

Here are examples of these models. You know the first one.

<u>Subtraction model 1</u>: <u>Take-away</u>

Suppose you have 29 objects and take 7 away. You will then have 29 + -7 or 29 - 7 objects left. This is, you will have 22 objects. In general,

<u>Take-away</u>
<u>model of</u>
<u>subtraction</u>:

| If b objects are taken from a objects, there are left <u>a - b</u> objects. |

<u>Subtraction model 2</u>: <u>Cutting off</u>

This model is like the take-away model. But it is more general. It applies to all **non**negative rational numbers.

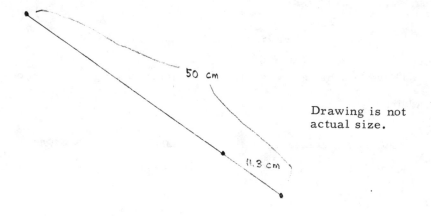

50 cm

11.3 cm

Drawing is not actual size.

146

Suppose you have a board with length 50 cm. If you cut off 11.3 cm, what will be left? The answer is 50 - 11.3 cm or 38.7 cm. In general:

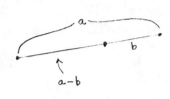

Cutting off
model of
subtraction:

| If something measuring b is cut off something measuring a (with same units), what is left has measure a - b. |

The "cutting off" model applies to more than length. It also can be applied to area.

Square ABCD at right has area 5·5 or 25. If the region ABE is cut off from the square, the area of the remaining portion is 25 - 6 or 19.

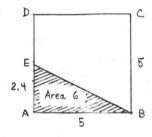

Other applications of the cutting off model are to volume, weight, money, and many other measurement situations. But the cutting off model does not apply to negative numbers. As in addition, there is a model related to slides.

Subtraction model 3: Directed distance.

Subtraction is often used to compare two numbers. Suppose you have $35.00 and want to buy an item costing $42.00. How much do you need?

Answer: 42.00 - 35.00 or 7.00

This problem can be pictured on a number line. The arrow stands for 7.

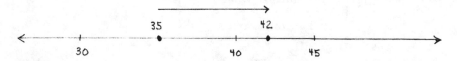

Thus 42-35 shows the direction and distance from 35 to 42.

In general,

| Directed distance model of subtraction: | a - b shows the distance and direction you must slide to get from b to a. |

$$60 - 100 = -40$$

$$-3 - -2 = -1$$

$$0 - 4 = -4$$

In each of the problems at left, b is bigger than a. The difference a - b is negative.

$$100 - 50 = 50$$

$$-2 - -30 = 28$$

$$\frac{1}{2} - -\frac{1}{2} = 1$$

Here b is smaller than a. The difference a - b is positive.

148

Thus subtraction tells how far apart two numbers are and which is larger. So subtraction is often used to <u>compare</u> two numbers.

Examples: Comparison by subtraction

1. Question: Yesterday's low temperature was -10°. Today's is 4°. What is the change in temperature?

Answer: 4 - -10 = 4 + 10 = 14. The temperature has gone up 14°.

This is easily seen on the number line.

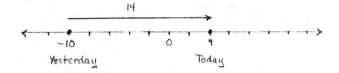

2. Here are scores of students on the same test given twice.

Student	September	June	June-September = "improvement"		
Carol	80	83	83 - 80	=	3
Darren	79	88	88 - 79	=	9
Ellie	75	74	74 - 75	=	-1
Frank	82	74	74 - 82	=	-8

The differences in the right column show which student improved most (Darren) and which least (Frank). Subtraction easily tells who improved, who didn't.

149

Questions covering the reading

1. In a schoolroom there are 24 students. If 8 leave, then how many are left?

2. What model of subtraction is found in Question 1?

3. If your allowance is A dollars a week and you spend S dollars on food, how much do you have left?

4. The area of the circle at right is 25 π . A square is cut out. Its area is 50. What is the area of the shaded region which remains?

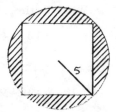

5. If AB = $\frac{2}{5}$ and AC = $\frac{3}{7}$, find the length of BC.

6. If BC = x and AC = n, what is the length of $\overline{AB}$?

7. What model of subtraction is found in Questions 3-6?

8. In 1960, Central H.S. had 2200 students. Now it has 2050 students. What has been the change in number of students from 1960 to now?

9. You have h dollars and want to buy something costing 100 dollars. How much do you need?

10. A person scores S on a test in September, J on a test in June. What is the amount of improvement?

11. What model of subtraction is found in Questions 8-10?

12-15. Give the change in the low temperature from yesterday to today if:

12. Yesterday it was 3°, today it is -4°.

13.   Yesterday it was -4°, today it is 3°.

14.   Yesterday it was 35°, today it is 31°.

15.   Yesterday it was 23°, today it is 28°.

16-19. Without actually calculating the answer, tell whether the answer is positive or negative.

16.   -17 - -16   17. 1.4 - -8.2

18.   469 - 398   19. -41.2 - 169

Questions testing understanding of the reading

1-4. Here are prices for some stocks on the New York Stock Exchange. Give the change in the price from September 22 to October 13, 1974.

	Stock	Sept. 22	Oct. 4	Oct. 13
1.	General Motors	40	35 3/8	36
2.	IBM	171 1/2	157 3/4	178
3.	Jones Laughlin	23 5/8	28	28
4.	US Steel	43 1/2	36	41 1/4

5-8. Give the net change in prices of these stocks (actual prices in week of Oct 7-11, 1974) on the Midwest Stock Exchange.

	Company	Oct. 4	Oct. 11
5.	Carson Pirie	8 3/8	8 5/8
6.	Hollys Inc.	3 5/8	3 1/4
7.	Piper Jaff	4 1/2	5
8.	Starr Best Grp	4	3 3/4

9-11. If I am y years old, how old is my sister who is:

9.   3 years younger.   10. 6 years younger.   11. x years older.

12.   A 3" square is cut off from an
        $8\frac{1}{2}$" by 11" sheet of typing
        paper. How much paper is left?

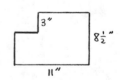

13. Suppose you weigh n kg now and you lose 2 kg in the next 7 days. What is your new weight?

14. A pollster wishes to predict the results of an election. He samples s people. Of these people, t are found to be too young to vote and another u people are not registered. How many people can the pollster use from the sample?

15. A board originally was 1 meter long. Now it is 75 cm long. How much was cut off of the board?

16. Angle ABD measures 53°.
    Angle CBD measures 36°.
    What is the measure of
    ∠ABC?

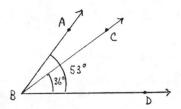

17. The population of Manhattan was 1,850,093 in 1900 and 1,539,233 in 1970 (according to the U.S. Census). What was the change in population?

18. (See Question 17.) If p people in Manhattan were under 60 years old in 1970, about how many were at least 60 years old?

19. If you have $100 in the bank and withdraw w dollars, how much is left?

20-23. Four people go to a health club. Weights are given for February 1st and March 1st. Give the change in weight for that month.

		Weight Feb. 1st	Weight Mar. 1st
20.	Jackie	100 kg	95 kg
21.	Ken	110 kg	114.5 kg
22.	Lulu	70.8 kg	67 kg
23.	Matt	83.4 kg	84.6 kg

24. (Refer to Questions 20-23.) If Mr. X weighed a kg on Feb 1st and b kg on Mar. 1st, what was his change in weight?

25-28. Suppose x is graphed on the number line. (a) How far is the given number from x? (b) Is the given number bigger or smaller than x?

25. x - .05   26. x - 47   27. x + $\frac{1}{100}$ 28. x - (-132)

Question for discussion

1. The following story appeared in Illinois Bell Telebriefs, Volume 37, No. 9, for October, 1975. A young boy called an operator for help, saying:

    "Operator, I'm sorry to bother you but I can't find the minus key on this telephone. Could you please call 555 take away 2368. I know it sounds impossible, but that's the way it's listed in the phone book, 555-2368."

    If you were the operator, how would you have answered the boy?

Lesson 7

A Statistic - The Mean

Six sisters seriously study statistics. They take a test and get the following scores: 91, 82, 96, 83, 90, and 91. How can their scores be described without listing them all?

Statistics can be used to describe these scores. A statistic is a number which is used to describe a collection of numbers. One statistic is the mean. (You probably call this statistic the average.)

> Let S_1, S_2, ..., S_n stand for n real
> numbers. Then the <u>mean</u> of these
> numbers is
> $$\frac{S_1 + S_2 + S_3 + \ldots + S_n}{n}$$

(You may wonder why the word "mean" is used instead of "average."
The term "average" is used in stock market averages, batting
averages, bowling averages, and these are all calculated differently!
The word "mean" - short for "arithmetic mean" - is used in only
one way and so will not be confusing.)

For the above scores, the mean score = $\dfrac{91 + 82 + 96 + 83 + 90 + 91}{6}$

$$= \frac{533}{6} = 88\frac{5}{6}$$

The mean score of the six sisters is $88\frac{5}{6}$.

You can check this calculation by adding up the differences
between each score and the mean. The sum should be 0.

Score	mean score	Score - mean
91	$88\frac{5}{6}$	$2\frac{1}{6}$
82	$88\frac{5}{6}$	$-6\frac{5}{6}$
96	$88\frac{5}{6}$	$7\frac{1}{6}$
83	$88\frac{5}{6}$	$-5\frac{5}{6}$
90	$88\frac{5}{6}$	$1\frac{1}{6}$
91	$88\frac{5}{6}$	$2\frac{1}{6}$

$$\text{sum} = 12\frac{4}{6} - 11\frac{10}{6} = 0$$

Since the sum is 0, $88\frac{5}{6}$ is the correct mean.

When the sum of some numbers is negative, their mean will also be negative. For example, suppose a person goes to the race track and his bets turn out as follows:

1st race: lost $5

2nd race: lost $2

3rd race: won $6.20

4th race: lost $4

5th race: won $3.20

Question: What did the bettor win or lose per race?

Answer: Add up the winnings and losses, then divide by 5.

$$\frac{-5 + {}^-2 + 6.20 + {}^-4 + 3.20}{5} = \frac{{}^-1.60}{5} = {}^-.32$$

The bettor lost 32¢ per race.

1. Define: statistic.   2. Define: mean.

3-10. Give the mean of each group of numbers.

3. 10, 20, 30   4. -6, -5, -1, 0, 0

5. $\frac{1}{2}, \frac{1}{2}, \frac{1}{2}, \frac{1}{2}$ 6. 5, -5, 4

7. a, b   8. x, y, z

9. S_1, S_2 10. S_1, S_2, S_3, S_4, S_5

11. The mean of 100, 200, and 240 is 180. Check this by sub-
    tracting the mean from each number and adding the differ-
    ences.

12. Repeat Question 11 to check whether 75 is the mean of 60,
    80, and 90.

13-16. True or False.

13. The mean of S_1, S_2, and S_3 is S_2.

14. The mean of three numbers a, b, and c may be larger than
    any of them.

15. If all numbers are positive, their mean is positive.

16. If all numbers are negative, their mean is negative.

Questions testing understanding of the reading

1-6. Calculate the mean of each group of data.

1. numbers found in counting problems: 1, 7, 21, 35, 35,
    21, 7, 1

C 2. S.A.T. scores: 600, 592, 586, 525, 525, 490, 462, 410,
    400, 390

3. numbers of children in some families: 0, 3, 2, 1, 1, 1,
    0, 5, 2, 4

4. measurements of body temperature during the day: 98.0,
    99.2, 98.3, 98.4

5. temperatures for Fairbanks, Alaska: $-12°$, $-20°$, $-18°$, $-17°$

6. scores in golf: 2 over par, 1 under par, 3 under par

7-10. Check each mean by subtracting it from each number and adding the differences.

7. 4, 5, 6, 8, 9   mean 6.4

8. -3, -1, -7, -1 mean -3

C 9. -1.3, 2.8, 4.6, -3.9, 0   mean .44

10. 5, 12   mean $8\frac{1}{2}$

11-12. Graph the numbers and their mean on a number line.

11. 46, 39, 41   12. -11, -5, 0, 4

13-14. Give the mean without calculating.

13. 3, 6, 9, 12, 15, 18, 21, 24, 27

14. -5, -5, -5, -4, -4, -4, -4, -5

Questions for discussion and exploration

1. In Question testing understanding #3, the mean number of children in a family was found to be 1.9. But it is impossible to have 1.9 children in a family. What does the 1.9 indicate?

2. It is possible to check a calculation of the mean by building a physical model.

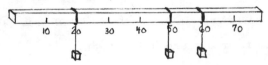

Take a meter stick or other long object and put a number line on it. If you want the mean of 20, 50, and 60, hang identical weights from these numbers (pictured above). Now try to balance the stick on your finger. The stick should balance on the mean. For the situation pictured, the balancing point would be at $43.\overline{3}$. (This is called the "center of gravity" property of the mean.)

157

Lesson 8

A Statistic - The Mean Absolute Deviation

When you are a junior or senior in high school, if you want to get into some colleges, you may have to take some exams called the S.A.T.'s (Scholastic Aptitude Tests). Scores on these exams may be as low as 200 or as high as 800.

Here are scores of ten top students from two schools.

School No. 1		School No. 2	
	750	800	
	742	792	
	736	786	
	725	725	
mean	725	725	mean
698	690	690	698
	662	662	
	660	610	
	650	600	
	640	590	

The sets of scores are different, but the means are 698. So the means don't show the difference. But No. 2's scores are more widely dispersed. One statistic which measures dispersion is called mean absolute deviation, abbreviated m.a.d. It is the average distance from a score to the mean.

We show how to calculate the m.a.d. using the scores of School No. 1.

Step 1. Calculate the mean. (It is 698.)

158

Step 2. Subtract the mean from each number. Step 3. If the difference in Step 2 is positive or zero, keep it. If the difference is negative, change it to its opposite.

750 - 698 = 52	52
742 - 698 = 44	44
736 - 698 = 38	38
725 - 698 = 27	27
725 - 698 = 27	27
690 - 698 = -8	8
662 - 698 = -36	36
660 - 698 = -38	38
650 - 698 = -48	48
640 - 698 = -58	58

Step 4. Take the mean of the numbers found in Step 3. It is 37.6, and this number is the m.a.d.

Now look at the four steps again. Steps 1, 2, and 4 are easy to describe in symbols. But Step 3 treats negative numbers differently from positive numbers. Never in this book have you had any operation which does this. The operation described in Step 3 is called <u>absolute valuing</u>.

Definition:

> The <u>absolute value</u> of x, written $|x|$, is defined in the following way.
>
> If x is negative, $|x| = -x$.
>
> If x is positive or zero, $|x| = x$.

Examples: Absolute valuing.

1. $|-3.2| = -(-3.2) = 3.2$ 2. $|97| = 97$

3. $|5 - 5| = |0| = 0$. The $|\ |$ sign has parentheses implied in it, so work inside the sign first.

4. $|4| - |-3| = 4 - 3 = 1$

There is a simple geometric interpretation for absolute value.

> The absolute value of a number is its distance from 0.

Opposites are the same distance from 0, so they have the same absolute value. That is, for any number a, $|a| = |-a|$.

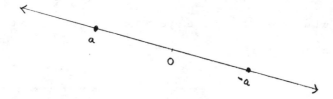

The symbol $|\ |$ is convenient for calculating the m.a.d. For the scores of School No. 2:

Step 1. The mean is 698 (calculated earlier).

160

Steps 2 and 3.

$$|800 - 698| = |102| = 102$$
$$|792 - 698| = |94| = 94$$
$$|786 - 698| = |88| = 88$$
$$|725 - 698| = |27| = 27$$
$$|725 - 698| = |27| = 8$$
$$|690 - 698| = |-8| = 8$$
$$|662 - 698| = |-36| = 36$$
$$|610 - 698| = |-88| = 88$$
$$|600 - 698| = |-98| = 98$$
$$|590 - 698| = |-108| = 108$$

Step 4.   Find the mean of the numbers in Step 3.   This mean is 67.6

Notice that Step 2 finds differences (<u>deviations</u>).   Step 3 calculates the <u>absolute</u> values of these.   Step 4 calculates the <u>mean</u> of these absolute deviations.   That is how the statistic gets its name.

In the two examples,

m. a. d. of School No. 1 = 37.6

m. a. d. of School No. 2 = 67.6

The larger m. a. d. indicates that in some way, the scores of School No. 2 are more dispersed.   They differ more from each other.

<u>Questions covering the reading</u>

1.   Give the definition of absolute value.

2-16.   Simplify:

2. $|46|$

3. $|-4|$

4. $|0|$

5. $-|-4|$

6. $|3| - |5|$

7. $-|12.4|$

8. $\left|-\frac{1}{2}\right|$

9. $|-3 + 8|$

10. $6 \cdot |-2|$

161

11. $|30 - 46|$ 12. $|46 - 30|$ 13. $|400 - 396.4|$

14. $|12 - 17.2|$ 15. $|5| - |-5|$ 16. $|-11 - 2|$

17. Give a geometric interpretation of absolute value.

18. Explain geometrically why $|a| = |-a|$ for any number a.

19. How is the mean absolute deviation calculated?

20. What is an abbreviation for "mean absolute deviation"?

21. Where does the name "mean absolute deviation" come from?

22-27. Calculate the mean absolute deviation for each sequence.

22. Bowling scores: 150, 145, 155

23. Golf scores: 90, 97, 104, 95

24. Low temperatures for Fairbanks in January: -12°, -20°, -18°, -17°

25. A number pattern: -5, -15, -25, -35

26. Heights of 13-yr-old quintuplets (in meters): 1.60, 1.60, 1.60, 1.60, 1.60

27. Number of accidents in a city for each day of a week: 27, 21, 30, 26, 29, 17, 15

28. What does a higher m.a.d. for some numbers indicate?

29. What are the S.A.T.'s?

30. What is the highest score possible on the S.A.T.'s? What is the lowest score possible?

31. Multiple Choice. The m.a.d. is a:

    (a) statistic which measures dispersion.
    (b) magazine.
    (c) description of some hatters.

162

1-4. The mean absolute deviation can be used to measure <u>consistency</u>. The more consistent set of scores is that with the smaller m.a.d. Calculate m.a.d.'s to answer these questions.

1. On her first 4 math tests, Freda scored 82, 71, 85, and 82. On his first 4 tests, Frank scored 84, 86, 70 and 80. Who was the more consistent student on these tests?

C 2. Two people work in a candy store filling orders for people. The boss decides to check their work by weighing five 1 kg boxes of candy from each worker. Here are the weights.

Worker A: 1.08, 1.04, 1.05, 1.01, 1.01

Worker B: 1.00, .98, 1.02, .96, 1.09

(1) Use the m.a.d. to determine who is more consistent.
(2) In your opinion, who is the better filler?

3. Which firm had the more consistent business?

	Firm A	Firm B
Profit 1st Year	-10,000	80,000
2nd	150,000	200,000
3rd	100,000	20,000
4th	200,000	140,000

C 4. Given are the normal temperatures for each month in San Francisco and Miami. In which city is the weather more consistent?

	J	F	M	A	M	J	J	A	S	O	N	D
SF	49	51	53	56	58	61	63	63	64	61	55	50
M	67	68	71	74	78	81	82	82	81	78	72	68

To whom might consistency of weather be important?

5-7. Choose the set of numbers which would probably have the smaller m.a.d.:

5. incomes of 10 union members from a factory; incomes of 10 management people from the same factory.

6. temperatures in one place on the moon for a month; temperatures in one place on Earth for a month.

163

7.   recovery times for 6 people from an appendectomy; recovery times for 6 basketball players from knee operations.

8-15. Suppose x can be any real number. Answer always (A), sometimes (S), or never (N).

8. $|x| = -|x|$ 9. $|x| \geqslant 0$

10. $|-x| \geqslant 0$ 11. $|x| < 0$

12. $|-x| < 0$ 13. $|x| > x$

14. $|x| \leqslant x$ 15. $|3 - x| = |x - 3|$

Lesson 9

Distance

Distance in mathematics applies to more than geography. In calculating the m. a. d. , the distance between a number and the mean is called the deviation.

deviation of score from mean = $|$score - mean$|$

For example, if a score is 75 and the mean score is 80. 3, the deviation is $|75 - 80.3|$ or $|-5.3|$ or 5.3. From a graph, you can see that 5.3 is the distance.

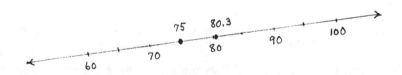

164

In other applications, distance is called <u>error</u>. Suppose a person guesses that there will be 250 students at a dance. Suppose 280 show up. The error is:

$$|280 - 250| = |30| = 30$$

In general, if you guess g students and s students show up, you will be off $|s - g|$. This is the formula for distance.

Definition:

> The <u>distance</u> between the numbers
> x and y is $|x - y|$.

Notice that x ‑ y and y ‑ x are the directed distances between x and y. But x ‑ y and y ‑ x are opposites. So $|x - y| = |y - x|$. The absolute value sign guarantees that $|x - y|$ is never negative. Thus $|x - y|$ is sometimes called the "undirected" distance between x and y.

Examples: Distance

1.   The distance between ‑3 and 2 is $|-3 - 2| = |-5| = 5.$

You could also get this answer by counting.

This is the difference between losing 3 yards and gaining 2 yards.

165

2. The distance between 47 and 46 = $\left|47 - 46\right|$ = $\left|1\right|$ = 1.
    You could also switch the numbers in this formula.
    The distance between 46 and 47 = $\left|46 - 47\right|$ = $\left|-1\right|$ = 1.

3. If two scores x and y differ by .01, then $\left|x - y\right|$ = .01
    The reason for using absolute value here is that you don't
    have to know which score is larger.

4. Suppose you estimate the age of a tree to be 40 years.   The
    tree is actually y years old.   How far off is your estimate?
    $$\text{Answer:} \quad \left|40 - y\right|$$
    You could answer $\left|y - 40\right|$ and still be correct.
    $\left|y - 40\right|$ = $\left|40 - y\right|$.

5. Suppose in Example 4 that you know your estimate is accurate
    to within 5 years.   Then
    $$\left|40 - y\right| \leqslant 5$$

Questions covering the reading

1. The distance between two numbers x and y is _____.
2-10. Give the distance between the numbers.

2. 10.1, 10.15   3. -6, -5   4. 0, -46.2

5. x, 3   6. 111.3, 110.8   7. 110.8, 111.3

8. $\frac{1}{3}, \frac{1}{4}$ 9. x, 5   10. 146, M

11. How is "distance" different from "directed distance"?

12. Which words can mean the same thing as "distance"?
   (a) error   (b) direction
   (c) comparison   (d) deviation

13. A cow weighs w. You estimate the weight to be 850. What is the error in your estimate?

14. It is estimated that the Chicago Bears will lose their next football game by 9 points. Instead they win by 3 points. What was the error in the estimate?

15. It is estimated that the Los Angeles Rams will win their next football game by E points. They actually win by A points. By how much was the estimate off?

16. You want a board of length ℓ . You find a board of length f. How far off are the boards lengths from each other?

17-18. True or False? Explain your answer.

17. $x - y = y - x$ 18. $|x - y| = |y - x|$

Questions applying the reading

1-6. <u>Distances along highways</u>. Interstate 10 (I 10) stretches from Florida to California. Here is a rough map. Mileages between some cities are given.

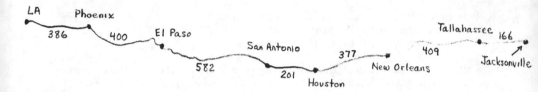

The map can be straightened. Someone who lives in Houston would coordinatize the map so that Houston has coordinate 0. Numbers now indicate mileage east or west from Houston. West is negative.

Use the definition of distance to find the distance between the given cities. Check your answer by adding mileages from the rough map.

1. Phoenix and Jacksonville

2. New Orleans and Jacksonville

3. Los Angeles and El Paso

4. Phoenix and Houston

5. San Antonio and Houston

6. Los Angeles and Tallahassee

168

7-10. <u>Closeness</u>. For these questions only, call a and b "close" if $|a - b| < .01$. Are a and b close?

7.   a = 2.54, b = 2.53   8.   a = -8.346, b = -8.355

9. $a = \dfrac{1}{11}$, $b = \dfrac{1}{10}$ 10.   a = 4,010,000, b = -4,010,000

11-13. Each situation can be written in the form $|x - y| = z$. For each sentence, tell what x, y, and z could be.

11.   The difference between the scores is 45.

12.   He and his sister are 3 years apart in age.

13.   The number n is .01 away from the number 6.

14-17. Each situation can be written in the form $|a - b| \leq c$. For each sentence, tell what a, b, and c could be.

14.   Those two mice were born within 3 days of each other.

15.   50% of people polled like the German sausage called bratwurst. The poll is almost always accurate to within 4%. The actual percentage of people liking bratwurst is p.

16.   I was travelling within 2 miles per hour of the 55 mph speed limit.

17.   That estimate of the number of people in the city was within 10,000 of the census figures.

18.   Describe the entire following situation in <u>one</u> sentence. (1) Doug estimates that D seniors in the school own cars. (2) Karen estimates that K seniors own cars. (3) A count is made and it is found that 55 seniors own cars. (4) Doug's estimate is better than Karen's.

19-22. <u>Time</u>. Years A.D. and B.C. can be placed on a number line called a <u>time line</u>. Then the time between two events can be calculated by using the definition of distance. This time may be off by as much as 2 years because (1) there was no year 0 and (2) each year itself covers a time period.

19.   According to the Old Testament and historical sources the first temple in Jerusalem was built by Solomon in about 970 B.C. and destroyed by Nebuchadnezzar in 597 B.C. How long did the temple stand?

169

20. The city of Alexandria, Egypt, was founded in 331 B.C. by Alexander the Great. How old is that city now?

21. It is now estimated that Jesus was born between 8 and 4 B.C. If Jesus was crucified in 29 A.D., how old was he then?

22. It is traditional to say that Islam was founded in 622 A.D., the date of Mohammed's flight from Mecca to Medina. How old is the Islamic religion? (Your answer will approximate the year this year is in the Islamic calendar.)

Question for discussion

1. Interstate highways (as in Question 1, p. 168) are numbered in a definite manner. What is this manner?

Chapter Summary

Let a and b be the numbers of elements in two sets which have no elements in common. How many elements are there altogether? Let a and b be the lengths of two segments. How long is the segment formed by joining the two segments? Let a and b each represent the direction and length of a slide. What number stands for the result of one of the slides followed by the other?

In all cases, the answer is $a + b$. The questions illustrate the three types of real situations which lead to the operation called addition. These situation-types are called models for addition.

Subtraction is defined in terms of addition: $a - b = a + {}^-b$. So it is natural that there is a model for subtraction related to each model for addition.

Addition model	Subtraction model
union	take-away
joining	cutting off
slide	directed distance

The models help in seeing some properties of the operations. For example, segments may be joined in any order and still would have the same total length. This idea is found in the <u>assemblage</u> <u>property</u>. The models also help get answers to addition and subtraction problems. For example, the slide model makes it obvious that the sum of two negative numbers is negative.

A <u>statistic</u> is a number which describes a collection of numbers. Two statistics are the <u>mean</u> (or <u>average</u>) and the <u>mean</u> <u>absolute deviation</u> (or m. a. d.). The mean is the sum of the terms divided by the number of terms. Calculation of the m. a. d. is helped by knowing <u>absolute value</u>. The absolute value of x, $|x|$, is the distance of x from zero. The <u>distance</u> between two numbers a and b is $|a - b|$. Distance is sometimes called <u>error</u> or <u>deviation</u>. The m. a. d. of a set of numbers is the mean distance of an element of the set from the mean.

Chapter 4

MULTIPLICATION

Lesson 1

The Repeated Addition Model for Multiplication

In the last chapter you studied 3 models each for addition
and subtraction.   Each model is related to a use of numbers.

Use of numbers	Addition	Subtraction
counting	union	take-away
measuring	joining	cutting off
comparison	slide	directed distance

In this chapter you will study and apply four basic models for
multiplication. There are models corresponding to the same three
uses of numbers.   The names for these models are listed here.
Underneath the name is the lesson where that model is explained.

Use of numbers	Multiplication
counting	ordered pair (Lesson 2)
measuring	area (Lesson 3)
comparison	size change (Lesson 5)

But we begin this chapter with a model for multiplication unrelated to a use of numbers. The model is suggested by the following question.

Question: If one can of peas costs 39¢, how much do 5 cans cost?

Answer: A cash register would add: $.39 + .39 + .39 + .39 + .39$

You would probably multiply: $5 \cdot .39 = \$1.95$

A number which is added is called a <u>term</u>. If the terms are all alike, you can always do a multiplication instead of adding them. Conversely, every multiplication by a positive integer can be done by repeated addition. You may have learned to multiply this way.

<u>Repeated Addition</u> <u>Model</u> for <u>Multiplication</u>:	If n is a positive integer, then $na = \underbrace{a + a + a + \ldots + a}_{n \text{ terms}}$.

That is,

$$1a = a$$

$$2a = a + a$$

$$3a = a + a + a$$

etc.

Because of this model, multiplication can be used to shorten some addition problems. Here are the scores of 25 students on a 9-question quiz given by the author. What was the mean score?

9 5 6 9 9 9 6 5 9 6 8 7 7 7 8 6 6 5 5 5 8 9 6 5 6

To find the mean, we need to add the scores. So we group them.

6 students scored 5.	$6 \cdot 5 = 30$
7 students scored 6.	$7 \cdot 6 = 42$
3 students scored 7.	$3 \cdot 7 = 21$
3 students scored 8.	$3 \cdot 8 = 24$
6 students scored 9.	$6 \cdot 9 = \underline{54}$
	171

The sum of the 25 scores is 171. So the mean score is $\frac{171}{25}$ or 6.84.

Recall that the numbers 6, 7, 3, 3 and 6 are the frequencies of the events "scoring 5," "scoring 6," etc. If you know the frequencies of the scores, you don't have to add 25 scores. Repeated additions can be replaced by multiplication.

The repeated addition model has some weaknesses. It does not apply to the multiplication of fractions. (It is difficult, if not impossible, to think of $\frac{2}{3} \cdot \frac{3}{7}$ as being repeated addition.) It does not apply to multiplication of negative numbers. This is why we need other models for multiplication.

Questions covering the reading

1.   What three models for multiplication will be discussed later in this chapter?

2.   What model for multiplication is given in this lesson?

3.   Give an example of the repeated addition model.

4.   What is a <u>term</u>?

5-8.   Simplify:

5. $y + y + y + y + y + y + y$ 6. $1.987 + 1.987 + 1.987 + 1.987$

7. $\underbrace{\frac{1}{3} + \frac{1}{3} + \ldots + \frac{1}{3}}_{100 \text{ terms}}$ 8. $\underbrace{a + a + \ldots + a}_{n \text{ terms}}$

175

9-18. From the given information, make up a question which can be answered either by repeated addition or by multiplication. Show the additions and the multiplication.

9.   You buy 100 13¢ stamps.

10.   You buy n 13¢ stamps.

11.   You buy n stamps at c cents each.

12.   A basketball arena seating 4500 people is sold out for 7 games in a row.

13.   A hockey arena seating p people is sold out for 7 games in a row.

14.   A stadium seating p people is sold out for g games in a row.

15.   A classroom contains 6 rows with 7 seats in each row.

16.   An auditorium contains 30 rows with s seats in each row.

17.   You and ten other students each collect 250 aluminum cans for recycling.

18.   You and ten other students each collect x aluminum cans for recycling.

19.   You score 96 on 3 tests and 91 on 2 others. What is your mean score?

20.   In one store you buy the last 2 bags of sugar and pay $2.49 a bag. In a second store, you buy 4 bags of sugar for $2.65 per bag. What is your mean purchase price?

21.   A diver receives 4 scores of 6.5 and 1 score of 6. What is the mean score?

22.   A gymnast receives 3 scores of 8.2, 2 scores of 8.4 and 1 score of 8.5. What is the mean score?

23-24. 20 scores are given. Calculate frequencies and use these to help find the mean of the scores.

23.   5 8 8 6 5 5 7 8 7 9 8 5 6 6 6 8 7 8 7 5

24.   0 1 1 0 2 2 1 0 1 1 0 1 2 2 1 0 1 1 0 2

Questions testing understanding of the reading

1-4. Do each problem in your head.

1. $19 + 19 + 19 + 19 + 19 + 19 + 19 + 19 + 19 + 19$

2. $\underbrace{12 + 12 + \ldots + 12}_{40 \text{ terms}} + \underbrace{12 + 12 + \ldots + 12}_{60 \text{ terms}}$

3. $1 \cdot 15 + 9 \cdot 15$ 4. $98 \cdot 4 + 2 \cdot 4$

5-8. Simplify:

5. $x + x + y + y + x$ 6. $a + 18 + a + a + a + 18 + 18 + 18$

7. $B + 9 + B + 9 + B + 9$ 8. $d + c + d + d + c + d$

9-14. Give the average of:

9. 30 scores of 90 and 20 scores of 95.

10. 15 scores of 6, 10 scores of 7, and 5 scores of 8.

11. m scores of 100 and n scores of 80.

12. t scores of 20, u scores of 30, and v scores of 40.

13. 3 scores of s_1 and 4 scores of s_2.

14. 1 score of s and 9 scores of t.

15. For a rock concert, $3 and $5 seats are available. If t seats
are sold at $3 and f seats of $5, how much money will be
collected?

16. A shoe store owner buys 1000 pairs of shoes at a cost c_1 per
pair and 500 pairs at cost c_2 per pair. (a) What is the total
cost? (b) What is the average cost?

17. On a 5-question quiz, 50 students scored as follows.

score:	0	1	2	3	4	5
frequency:	1	3	5	15	21	5

What is the mean of the scores?

18. You throw a die 100 times with the following results.

Lands on:	1	2	3	4	5	6
Frequency:	13	19	18	21	13	16

What is the mean of the 100 numbers?

19-20. Suppose you have c classes a day. Suppose there were 18 school days in September and 23 in October.

19. How many classes did you have in September?

20. How many classes did you have in September and October together?

21-22. Use repeated addition to verify:

21. $4 \cdot \frac{2}{3} = \frac{8}{3}$

22. $6 \cdot \frac{3}{2} = 9$

Questions for discussion and exploration

1. There is a <u>repeated subtraction</u> model of division. It is applied every time you do "long division." Look at the following long division problem.

$$
\begin{array}{r}
103 \\
38 \overline{\smash{)}3914} \\
\underline{38} \\
114 \\
\underline{114}
\end{array}
$$

Certainly subtraction is in this process. But where is the <u>repeated</u> subtraction?

2. On Halloween, a family stocked up with 200 pieces of gum to give out to "Trick or Treaters." If they gave out 3 pieces of gum to each child, how many children could they treat in this way? Explain how this is an application of the repeated subtraction model of division.

3-8. You could think of paying 39¢ as ⁻.39. Then if you buy 5 items for 39¢ each, the total is

$$⁻.39 + ⁻.39 + ⁻.39 + ⁻.39 + ⁻.39 = ⁻1.95$$

That is, $\qquad 5 \cdot ⁻.39 = ⁻1.95$

In this way, the repeated addition model applies to multiplication of a negative number by a positive integer. Use this idea to answer the questions.

3. Show that $4 \cdot ⁻3 = ⁻12$.

4. Show that $2 \cdot ⁻14 = ⁻28$

5. Calculate $10 \cdot ⁻\frac{1}{2}$ 6. Calculate $7 \cdot ⁻31$

7. If $⁻9 + ⁻9 + ⁻9 + ⁻9 + ⁻9 = n \cdot ⁻9$, what is n?

8. If $⁻2.3 + ⁻2.3 = m \cdot ⁻2.3$, what is m?

Lesson 2

The Ordered Pair Model for Multiplication

Here is a counting problem whose answer is given by multiplication.

Question: Four boys and three girls are on the freshman tennis team. How many mixed doubles teams are possible?

Solution: Call the boys A, B, C, and D. Call the girls X, Y, and Z.

179

Method 1:   Each boy can be paired with any of
three girls.

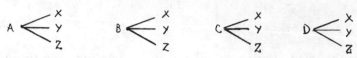

There are $3 + 3 + 3 + 3$ or $4 \cdot 3$ or 12

pairs possible.   This kind of picture is

called a "tree diagram," because the

parts of the picture look like branches.

Method 2:   Write the boys vertically, the girls
horizontally.   The points indicate the
pairs.

A   .(A, X)   .(A, Y)   .(A, Z)

B   .(B, X)   .(B, Y)   .(B, Z)

C   .(C, X)   .(C, Y)   .(C, Z)

D   .(D, X)   .(D, Y)   .(D, Z)

      X   Y   Z

The picture in Method 2 resembles the coordinates used in

graphing.   It suggests both the name and the description for the

counting model of multiplication.

If the first coordinate of an ordered pair can be any of x elements and the second coordinate can be any of y elements, then xy ordered pairs are possible.

Choosing coordinates is like filling blanks. Suppose the left blank can be filled in 7 ways. If the right blank can be filled in 5 ways, there are 7 • 5 or 35 pairs possible.

$\overline{\hspace{2cm}}$ 7 fillers $\overline{\hspace{2cm}}$ 5 fillers   35 possible pairs

Every ordered pair associates a first coordinate with a second coordinate. So you can multiply 7 by 5 by connecting points.

first
coordinate

second
coordinate

These are 35 segments connecting the 7 points at left to the 5 points at right. One of the segments is drawn.

181

Thus the ordered pair model for multiplication can be pictured in many ways:

tree diagram

horizontal-vertical graphing

filling blanks

connecting points

Which picture is best? It depends on the situation. In this book, the filling-blanks model will be used most often.

Questions covering the reading

1. How many different segments are there which connect points in Row 1 to points in Row 2.

    Row 1 . . . .

    Row 2 . . .

2. Verify your answer to Question 1 by tracing the rows and drawing all of the possible segments.

3. Describe the ordered pair model for multiplication.

4-6. Picture the fact that 3 • 5 = 15 by:

4. "connecting dots."

5. drawing a tree diagram.

6. drawing a horizontal-vertical ordered pair picture.

7-9. Repeat Exercises 4-6 for the problem of multiplying 5 by 4.

10. A classroom has six rows, each with seven seats. How many seats are in the classroom?

11. Seven girls and five boys try out for the lead parts in a play. How many different boy-girl couples could be formed?

12. A classroom has r rows, each with 7 seats. How many seats are in the room?

13. Suppose you own 10 records and 10 tapes that you are tired of. If you want to give away 1 record and 1 tape, how many different record-tape combinations do you have to choose from?

14. Repeat Question 11 if g girls and b boys try out for the parts in the play.

15. Repeat Question 13 if you are tired of 12 of your records and t of your tapes.

Questions testing understanding of the reading

1. Show that this problem fits the ordered pair model for multiplication: Four boys and no girls are on the freshman tennis team. How many mixed doubles teams are possible?

2. Multiple Choice. In a section of football stands there are 30 rows and each row contains 18 seats. The number of seats in this section is then closest to:

   (a) 100   (b) 300   (c) 500   (d) 2400

3. Jesse Jackson's initials are JJ. Gerald Ford's are GF. If a company wanted to make handkerchiefs with every possible pair of initials, how many types of handkerchiefs would they have to make?

4. Does this problem fit the multiplication counting model? "A person owns one shirt and 7 pairs of slacks. How many shirt-slack outfits are possible?

5. (a) If you had to learn all the multiplication "facts" from 1 times 1 to 12 times 12, how many facts did you have to learn? (b) How many multiplication facts did you have to learn if you also knew that multiplication is commutative (that is, $3 \cdot 4 = 4 \cdot 3$, etc.)?

6-8. These problems do not exactly fit the counting model. It may help to use tree diagrams.

6. Each of five cities is connected to the others by non-stop plane flights. How many plane flights are needed?

7. There are 8 teams in a basketball league. How many games are needed if each team is to play every other team on its home court?

8. There are n teams in a basketball league. How many games are needed if each team is to play every other team on its home court?

9-11. Suppose (x, y) is an ordered pair. How many ordered pairs are there like this if x and y are each:

9. 1-digit positive integers.

10. 2-digit positive integers.

11. 3-digit positive integers.

12. Multiple Choice. A high school contains 590 freshmen and 520 seniors. The number of possible freshmen-senior pairs is closest to:

(a) 550   (b) 1100   (c) 2200   (d) 300,000

13. Repeat Question 12 for a school with 30 freshmen and 40 seniors

Skill review

If you had trouble with the arithmetic in the above problems,

or if you are unsure of how accurately you can multiply, you may

need a skill review.

Directions: Try the first 5 problems in each group. Then look at the answers. (They are upside down on the page 186.) If you have 4 or 5 right, go on to the next type of problem. If you get 1, 2, or 3 correct, then you need more practice. Then you should try the last 5 problems. Then look at the answers to these. If you do not have 4 or 5 right, ask your teacher for help.

1-10. Multiply the given nonnegative integers.

1. 17, 9   2. 1, 13   3. 26, 7   4. 317, 23

5. 82, 994   6. 7, 12   7. 15, 15   8. 39, 8

9. 76, 212 10. 98, 103

11-20. Multiply in your head.

11. 20, 3   12. 10, 34   13. 100, 9   14. 60, 90

15. 50000, $\frac{1}{2}$ 16. 70, 5   17. 10, 476   18. 12, 100

19. 50, 80   20. $\frac{1}{5}$, 4000

21-30. Multiply the given positive rational numbers.

21. $\frac{1}{2}, \frac{1}{3}$ 22. $\frac{4}{9}, \frac{5}{9}$ 23. $\frac{11}{5}, \frac{10}{22}$ 24. 6, $\frac{2}{3}$

25. $\frac{5}{8}$, 20   26. $\frac{7}{3}, \frac{7}{2}$ 27. $\frac{5}{11}, \frac{4}{11}$ 28. $\frac{3}{14}, \frac{7}{6}$

29. 12, $\frac{5}{4}$ 30. 30, $\frac{7}{12}$

31-40. Multiply the given decimals.

31. 1.4, 1.3   32. .14, .13   33. 681.2, .03

34. .46, 32   35. 8.6, 8.6   36. .108, .302

37. .5, .6   38. 20.8, .416   39. 46.3, 6

40. 110, .2

41-50. Which product is larger? (You should be able to do these
problems without actually finding the products.)

41. $1 \cdot 0$ or $1 \cdot \frac{1}{2}$ 42. $20 \cdot 18$ or $19 \cdot 17$

43. $86.2 \cdot 49$ or $85.9 \cdot 49$ 44. $\frac{3}{4} \cdot 50$ or $\frac{2}{5} \cdot 50$

45. $\frac{1}{2} \cdot \frac{1}{4}$ or $.5 \cdot .25$ 46. $\frac{3}{4} \cdot 50$ or $\frac{3}{5} \cdot 50$

47. $0 \cdot 17$ or $1 \cdot 15$ 48. $80 \cdot 0.2$ or $80 \cdot 0.3$

49. $11.2 \cdot \frac{1}{100}$ or $11.2 \cdot 0$ 50. $8 \cdot 15$ or $7 \cdot 14$

51-60. Multiply all of the given numbers.

51. 2, 3, 4   52. 17, 15, 0   53. 6, 0, 60

54. 10, 100, 50   55. 1.1, 2.2, 3.3   56. 8, 8, 9

57. $\frac{2}{3}, \frac{3}{4}, \frac{4}{5}$ 58. 60, 30, 20   59. $\frac{1}{2}, \frac{1}{2}, \frac{1}{2}$

60. 2.1, 2.2, 2.3

185

Answers to skill review

1. 153 2. 143 3. 182 4. 7291 5. 81,508 6. 84 7. 225

8. 332 9. 16112 10. 10094 11. 60 12. 340 13. 900

14. 5400 15. 25000 16. 350 17. 4760 18. 1200 19. 4000

20. 800 21. 1/6 22. 20/81 23. 1 24. 4 25. 25/2 (or 50/4

or 100/8) 26. 49/6 27. 20/121 28. 1/4 29. 15 30. 35/2 (or

210/12 or 105/6) 31. 1.82 32. .0182 33. 20.436 34. 14.72

35. 73.96 36. .032616 37. .3 38. 8.6528 39. 277.8 40. 22

41. second 42. first 43. first 44. first 45. equal 46. first

47. second 48. second 49. first 50. first 51. 24 52. 0

53. 0 54. 50,000 55. 7.986 56. 576 57. 2/5 58. 36,000

59. 1/8 60. 10.626

Lesson 3

The Area Model for Multiplication

The ordered pair model for multiplication applies only to nonnegative integers. A more general model applies to all non-negative reals. It is well known to you.

<table>
<tr><td>Area model
for multiplication:</td><td>If a rectangle has dimensions x and
y, then its area is xy.</td></tr>
</table>

Examples of the area model show how applicable it is. If you forget how to multiply, the area model can often get you an answer.

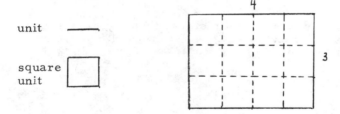

unit   ———

square
unit

4

3

1. Positive integers: 3 • 4 = _____

   Counting the square units
   shows the area to be 12.

187

2. Fractions: $\frac{3}{2} \cdot \frac{2}{5} =$ _____

The unit area has been divided
into 2·5 or 10 parts. So each
little rectangle has area 1/10
of a square unit. There are
3·2 or 6 parts in the dark
rectangle.
So its area is 6/10.

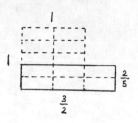

3. Decimals can be converted
into fractions so you can
get products of decimals
by the same kind of process.

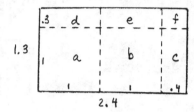

1.3 · 2.4

$$= a + b + c + d + e + f$$
$$= 1 + 1 + .4 + .3 + .3 + .12$$
$$= 3.12$$

In the area model, the product xy can be in <u>different units</u>

than x or y.

4. If x and y are in centimeters, then xy is in <u>square centi-
meters</u>. For example, if x = 2 cm and y = 3.4 cm, then
xy = 6.8 sq cm.

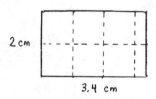

2 cm · 3.4 cm = 6.8 sq cm

5. Suppose 5 men work $1\frac{1}{2}$ hours a piece on a job. We say that the job takes $5 \cdot 1\frac{1}{2}$ or $7\frac{1}{2}$ man-hours.

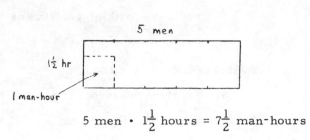

$$5 \text{ men} \cdot 1\frac{1}{2} \text{ hours} = 7\frac{1}{2} \text{ man-hours}$$

6. If a 100-watt light bulb burns for 50 hours, then $50 \cdot 100$ or 5000 watt-hours of energy have been used. The watt-hour is a metric unit. Since 5000 watts = 5 kilowatts,

$$100 \text{ watts} \cdot 50 \text{ hours} = 5000 \text{ watt-hours}$$
$$= 5 \text{ kilowatt-hours}$$

Most electric bills in the U.S. give the energy used in kilowatt-hours.

Examples 4-6 show that it is helpful at times to think of "multiplying" the <u>units</u> of measurement.

Among the things which the area model shows are the following:

1. Multiplication is commutative. (Switching the dimensions of a rectangle will not change its area.)

2. Suppose x_1 is close to x_2 and y_1 is close to y_2. Then $x_1 y_1$ will be close $x_2 y_2$. (The rectangle with dimensions x_1 and y_1 will look very much like the rectangle with dimensions x_2 and y_2.)

189

3. Larger positive numbers yield larger products. (A

rectangle with bigger dimensions will have a larger area.)

Questions covering the reading

1. What is the area model of multiplication?

2-3. What multiplication problem is pictured?

2.

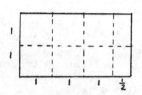

3.

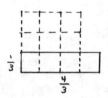

4-7. <u>Accurately</u> draw a rectangle with the given dimensions. Show that its area is the product of its dimensions.

4. 4 cm and 5 cm

5. 4.2 cm and 5 cm

6. $1\frac{5}{8}$ in. and 3 in.

7. $\frac{2}{3}$ and $\frac{3}{5}$ (pick your own unit)

8. Make an accurate drawing of rectangles to show that $5.1 \cdot 4$ is larger than $5 \cdot 4$.

9. How do rectangles show that multiplication is commutative?

10. Suppose 30 secretaries each work 16 hours to get out a large mailing. How many man-hours were used? (Note: The name "man-hour" is commonly used even if some or all of the workers are women.)

11-12. A 150-watt bulb is left burning for 2 days.

11. How many watt-hours of energy have been used?

12. How many kilowatt-hours of energy have been used?

13-14. Repeat Questions 11-12 for a 60-watt bulb which is left on for a week.

190

Questions testing understanding of the reading

1. Show that $\frac{2}{3} \cdot \frac{3}{2} = 1$ by drawing a rectangle with appropriate dimensions.

2. Use rectangles to show that $\frac{1}{3} \cdot \frac{1}{4} = \frac{1}{12}$.

3. Draw a rectangle. Call its dimensions ℓ and w. Draw a second rectangle with dimensions 3ℓ and 3w. How do the areas compare?

4. Calculate the area of a rectangle with sides 5 and x. **Now** multiply the lengths of the sides by 4. What happens to the area?

5-8. Monthly electric rates for a residence in Illinois (as of April, 1974) were as follows (quoted from Commonwealth Edison pamphlet):

> Monthly charge $0.95
> 1st 100 kwhrs of use   .0385 per kwhr
> Next 225 kwhrs of use   .0282 per kwhr
> Over 325 kwhrs of use   .0279 per kwhr

(The abbreviation "kwhr" stands for "kilowatt-hour.") Tax is added to this and adjustments are made if the price of energy to Commonwealth Edison changes.

What would be the monthly charge for each amount of energy?

5. 100 kwhrs   6. 200 kwhrs

7. 400 kwhrs   8. 700 kwhrs

9-10. An old unit of work is the foot-pound. One foot-pound is the amount of work done by a force of one pound moving through a distance of one foot.

9. How many foot-pounds of power are needed to move 1000 pounds a distance of 50 feet?

10. Generalize Question 9 and its answer.

11-12. (adapted from a problem in <u>Practical Problems in Mathe-</u>
<u>matics for Carpenters</u> by Jack Luy, Delmar Publ., 1973)
In the site plan illustrated:

11. What is the area in square feet which will need to be seeded to establish a lawn?

12. What is the area in square yards which will need to be seeded to establish a lawn?

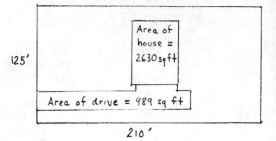

125'

Area of house = 2630 sq ft

Area of drive = 989 sq ft

210'

13-15. Give the area of the rectangle with dimensions:

12. 4 and t   14. x and 2x   15. n + 1 and n + 2

16-17. Use these segments of length a, b, and c.

a   b   c

16. Draw a rectangle whose length is a + c and whose width is 5 cm. Divide this rectangle into two small rectangles to show that 5(a + c) = 5a + 5c.

17. Draw a rectangle whose length is a + b and whose width is c. Divide this rectangle into two small rectangles to show that (a + b)c = ac + bc.

18-21. The four vertices of a rectangle are given. (1) Graph the rectangle using the same scale on each axis. (2) Find the area of the rectangle.

18. (8, 0), (9, 0), (8, 12), (9, 12)

19. (7, -11), (-4, -11), (-4, -6), and (7, -6)

20. (x, 0), (x + 1, 0), (x + 1, 10), and (x, 10)

21. (This one is harder.) (1, 1), (5, 6), (6, -3), (10, 2)

22. Does the formula $A = \ell w$ work for a rectangle like the one drawn above? Explain your answer.

Lesson 4

The Assemblage Properties of Multiplication

One year, about half of the students in the author's class thought that these two multiplication problems would give different answers.

$$
\begin{array}{c}
38.2 \\
\underline{97}
\end{array}
\qquad\qquad
\begin{array}{c}
97 \\
\underline{38.2}
\end{array}
$$

These students did not understand that multiplication is <u>commutative</u>. The answers to the above problems are the same. The area model makes this obvious. Look at the rectangles below. The area of the left rectangle is $38.2 \cdot 97$. The area of the right rectangle is

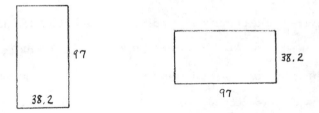

$97 \cdot 38.2$. The areas are the same. So $97 \cdot 38.2 = 38.2 \cdot 97$. More generally, $\ell w = w \ell$.

With three numbers to be multiplied:

$$35 \cdot 8 \cdot 41$$

Will you get the same answer if you multiply 8 by 41 first as you would get by following order of operations? It is easy to find out.

$35 \cdot 8 \cdot 41$	$35 \cdot (8 \cdot 41)$
$= 240 \cdot 41$	$= 35 \cdot 288$
$= 9840$	$= 9840$

This is an instance of the <u>associative property of multiplication</u>. By extending the area model to 3 dimensions, it is possible to show that the associative property always holds. This extended area model is the volume model.

<u>Area model (extended to 3 dimensions)</u>: <u>Volume</u>

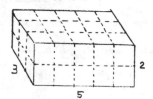

If a rectangular solid has dimensions x, y, and z, then its volume is xyz.

Pictured is a rectangular solid with height 2, length 5, and width 3. Its volume is $2 \cdot 3 \cdot 5$ or 30. You could look at this solid from a different angle. You could move it. This might cause the dimensions to be switched. But the volume would be unchanged. In this way, $2 \cdot 3 \cdot 5$ is equal to $2 \cdot 5 \cdot 3$, $3 \cdot 2 \cdot 5$, and any other reordering.

Volume is measured in <u>cubic</u> units.

—	☐	⬜
unit	square unit	cubic unit

If the dimensions of the rectangular solid had been 2 cm, 5 cm, and 3 cm, then the volume of the solid would be 30 cubic centimeters. (That is a way of saying that 30 centimeter cubes would fit in the solid.) This is abbreviated 30 cc. Blood and many other fluids are often measured in cc's.

Because the dimensions can be switched without affecting the volume, multiplication has an assemblage property.

<u>Assemblage property</u>
<u>of multiplication</u>:

Given some numbers to be multiplied, changing the order of the numbers or the order of multiplications does not affect the product.

The assemblage property guarantees that multiplication is both commutative and associative.

Examples: Applying the assemblage property

1.　　Multiply:　$5 \cdot 5 \cdot 5 \cdot 2 \cdot 2 \cdot 2$

　　　　Solution:　It is easier to reorder than to multiply from

　　　　　　　　　left to right.

$$= 5 \cdot 2 \cdot 5 \cdot 2 \cdot 5 \cdot 2$$

$$= 10 \cdot 10 \cdot 10$$

$$= 1000$$

2.　　Multiply:　$\frac{5}{6} \cdot \frac{1}{2} \cdot \frac{3}{4} \cdot \frac{12}{5} \cdot \frac{1}{3}$

　　　　Solution:　Reorder.

$$= \frac{5}{6} \cdot \frac{12}{5} \cdot \frac{1}{2} \cdot \frac{3}{4} \cdot \frac{1}{3}$$

$$= 2 \cdot \frac{1}{2} \cdot \frac{3}{4} \cdot \frac{1}{3} = \frac{3}{4} \cdot \frac{1}{3} = \frac{1}{4}$$

Questions covering the reading

1.　Multiply. Do both problems. Do not use a calculator.
Show all work.

$$\begin{array}{cc} 132.4 & 89.1 \\ \underline{89.1} & \underline{132.4} \end{array}$$

2.　Multiply. Do both problems. Do not use a calculator.
Show middle steps and all work.

$$(\frac{1}{2} \cdot \frac{4}{3}) \cdot \frac{3}{5} \qquad\qquad \frac{1}{2} \cdot (\frac{4}{3} \cdot \frac{3}{5})$$

3.　What property of multiplication is verified by Question 1?

4.　What property of multiplication is verified by Question 2?

5-10. Give the volume of a rectangular solid with dimensions:

5.　2 cm, 15 cm and 3 cm.　　　　6.　1.8, 3.9, and 4.6.

196

7. $\frac{3}{8}$ in., $\frac{1}{2}$ in., and $1\frac{1}{4}$ in.   8. 18, 39, and 46

9. a, b, and c   10. 2a, 3b, and c

11. What is a cubic centimeter?

12. What is the abbreviation for "cubic centimeter?"

13. What is the assemblage property of multiplication?

14. How does the volume model help to show that multiplication has the assemblage property?

15. Choose the most general property: assemblage, associative, commutative.

Questions testing understanding of the reading

1. If $4 \cdot 5 \cdot 6 \cdot 7 = 840$, what is $2 \cdot 7 \cdot 6 \cdot 5 \cdot 4$?

2-7. Simplify in as easy a way as you can.

2. $\frac{1}{2} \cdot \frac{3}{4} \cdot \frac{5}{6} \cdot \frac{7}{8} \cdot \frac{2}{3} \cdot \frac{4}{5} \cdot \frac{6}{7}$ 3. $25 \cdot 25 \cdot 25 \cdot 4 \cdot 4$

4. $1 \cdot 2 \cdot 3 \cdot 4 \cdot 5 \cdot 6 \cdot 7 \cdot 8 \cdot 9 \cdot 0$

5. $1 \cdot 2 \cdot 3 \cdot 4 \cdot 5 \cdot 6 \cdot 7 \cdot 8 \cdot 9 \cdot 10$

6. $2 \cdot 3 \cdot 4 \cdot 5 \cdot \frac{1}{2} \cdot \frac{1}{3} \cdot \frac{1}{4} \cdot \frac{1}{5}$

7. $3 \cdot 5 \cdot 4 \cdot 2 - 2 \cdot 4 \cdot 5 \cdot 3$

8-11. Use properties of addition and the repeated addition model of multiplication to help simplify.

8. xy - yx + xy + xy   9. abc + cba + bac + cba

10. mn + pq + nm + qp   11. vw - wv + wv

12-13. The metric units of mass and length are related in a very simple way. One cc of water weighs 1 gram. Give the weight of an aquarium filled with water if the aquarium is:

12. 30 cm high, 40 cm wide, and 18 cm deep.

13. twice as high, wide, and deep as in Question 12.

197

14-16. 1 cubic foot of water weighs about 62.4 pounds. How much will the water weigh if it fills:

14.   an aquarium 16" long, 8" deep, and 10" high.

15.   an aquarium whose dimensions are three times those in Question 14.

16.   a water bed which is 6" deep, 6' long, and $4\frac{1}{2}$' wide.

17.   You see a large figurine in a window and wonder what it weighs. There is a smaller similar figurine which is 1/3 as high, 1/3 as wide, and 1/3 as deep. You pick it up and estimate its weight to be 2 kg. What should you estimate for the weight of the large figurine?

Lesson 5

The Size Change Model for Multiplication

You have now seen three models for multiplication: repeated addition, ordered pair, and area. None of these models applies to negatives. The fourth and last model for multiplication is the size change model. This model can involve any real number, including negatives. It comes from the following kinds of questions. In each case, multiplying the two given numbers will help answer the question.

1.   There are about 21,000,000 Blacks in the U.S. and about 8% carry the sickle-cell trait. How many Blacks carry this trait? (Carrying the trait is not the same as having the disease.)

        Answer: about .08(21,000,000) or 1,680,000

2. Suppose you have a job which pays time-and-a-half for over-
    time. If you make $3.50 an hour, how much would you make
    per hour of overtime?

    Answer: 1 1/2 · 3.50 = 5.25, so you would make $5.25 per
    hour.

3. Doll's house furniture is often 1/16 the size of normal furni-
    ture. A typical chair can be 80 cm high. How high would
    the corresponding doll's furniture be?

    Answer: 1/16 · 80 = 5, so the doll's chair would be 5 cm high.

In Question 3 above, $\frac{1}{16}$ is a <u>scale factor</u>. A furniture maker
for dolls often would multiply dimensions of actual furniture by $\frac{1}{16}$.
This would give the dimensions of the doll's furniture. In Example
1, 8% is the scale factor. In Example 2, 1 1/2 is the scale factor.

Percentages are almost always scale factors. When you see
"Prices slashed 20%," you should know that multiplication is involved.
These segments picture what happens.

original price

amount of reduction

The cutting off model of subtraction tells you that the final price
would be p - .20p.

final price

If the original price were raised 30%, then the amount of increase
would be .30p. The joining model of addition would apply.

amount of increase

final price

199

The name of the size change model comes from the following application.

Each coordinate of the points of the larger figure has been multiplied by 2/3. The smaller figure results.

The size change model applies also to multiplication with negative numbers. This is the topic of the next lesson.

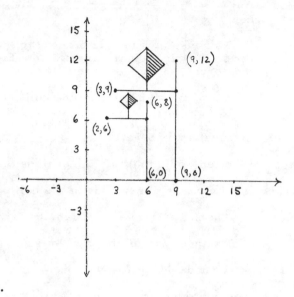

Questions covering the reading

1-4. About how many Blacks in the given city carry the sickle-cell trait?

1.   New York (estimated Black population 1, 700, 000)

2.   Fort Worth (estimated Black population 80, 000)

3.   San Diego (estimated Black population 53, 000)

4.   a city with an estimated Black population of B

5-8. At time-and-a-half for overtime, how much will you make per hour of overtime if your normal pay is:

5. $3.00/hr   6. $2/hr   7. $2.75/hr   8. $x/hr

9-12. Using the scale factor given in the lesson, what would be the height of doll-house furniture corresponding to:

9.   a table 1 meter high.   10.   a bed with height b.

11.   a lamp with height ℓ.   12.   a house 30 feet high.

13-15. Carefully graph the points (1, 0), (1, 3), and (2, 6). Connect the points to form a triangle. Then multiply all the coordinates by the given scale factor and graph the new triangle on the same graph.

13.   scale factor 3   14.   scale factor 2.5

15.   scale factor $\frac{1}{4}$

16.   What is the scale factor in Questions 1-4, p. 200 ?

17.   What is the scale factor in Questions 5-8, p. 200 ?

18-21. An item now costs $30. What will be the new cost of this item if the present cost is:

18.   raised 20%.   19.   lowered 20%.

20.   lowered 50%.   21.   raised 50%.

22.   Repeat Questions 18 and 20 for an item which costs d dollars.

23.   Repeat Questions 19 and 21 for an item which costs c dollars.

24.   Draw a bar graph showing original and final costs in Question 22.

25.   Draw a bar graph showing original and final costs in Question 23.

1.   What happens when the scale factor in multiplication is zero?

2-3. Suppose the cost of living doubles from time A to time B.
Then what cost x in time A will cost 2x in time B.

2.   Here is a bar graph of costs of items in time A.

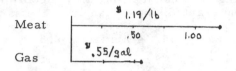

      Using the same scale, draw a bar graph of the costs of these
      items in time B.

3.   Question 2 fits the size change model. What is the scale
      factor?

4-6. Some estimate that 5% of the 210,000,000 people in the U.S.
are alcoholics.

4.   If this is true, how many people in the U.S. are alcoholics?

5.   Draw a bar graph with segments representing (a) the popula-
      tion and (b) the number of alcoholics in the U.S.

6.   Refer to Question 5. If the bar representing the population
      has length L, then the bar representing the alcoholics has
      length _____.

7-8. Each situation leads naturally to a size change model of multiplication. Name the scale factor.

7.   He is 48 years old and his son is half as old.

8.   She made $5000 last year. This year she will make 3 times
      as much.

9-10. Suppose a model plane is $\frac{1}{100}$ actual size.

9.   A plane wing of 30 m will be how long on the model?

10.   A part of length x cm on the plane will correspond to a piece
      of length _____ on the model.

202

11-12. Repeat Exercises 9-10 if the model is $\frac{3}{200}$ actual size.

13-14. Suppose you are looking through a microscope which magnifies 500 times.

13. An object .01 mm in length will seem to have what length in the microscope?

14. An object with actual length L will seem to have what length when viewed through the microscope?

Questions for discussion

1. Suppose you had a blueprint which you wanted to enlarge to $1\frac{1}{2}$ times its present size. Give some ways this might be done.

2. The population of Colorado Springs nearly doubled from 1960 to 1970. The graph below seems to show this. What is misleading about this graph?

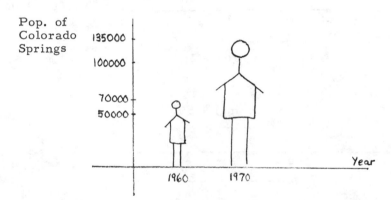

Lesson 6

Multiplication with Negative Numbers

It is customary in business to pay "time-and-a-half" for overtime. That is, if normal pay is x dollars an hour, overtime pay is $\frac{3}{2}x$ dollars an hour.

<u>normal</u>   <u>overtime</u>

x $\frac{3}{2}x$

Graphically:

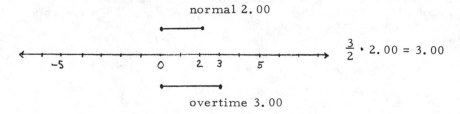

$\frac{3}{2} \cdot 2.00 = 3.00$

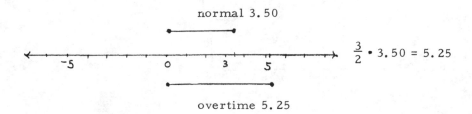

$\frac{3}{2} \cdot 3.50 = 5.25$

This is looking at it from the worker's point of view. But the boss is paying money <u>out.</u> For the boss, all pay is represented by <u>negative</u> numbers.

204

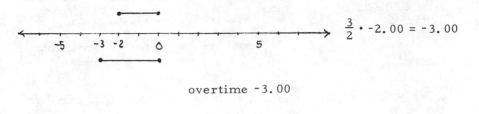

normal ⁻2.00

$$\frac{3}{2} \cdot {}^-2.00 = {}^-3.00$$

overtime ⁻3.00

normal ⁻3.50

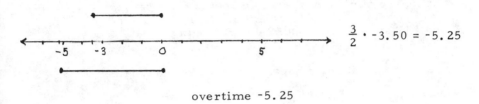

$$\frac{3}{2} \cdot {}^-3.50 = {}^-5.25$$

overtime ⁻5.25

Think of positive and negative as directions, like gain or loss. The pay example shows that multiplying by a positive number does not change directions.

multiplier		direction		direction
$\frac{3}{2}$	$\cdot$	gain	=	gain
$\frac{3}{2}$	$\cdot$	loss	=	loss

multiplier	direction	direction
positive	• positive	= positive
positive	• negative	= negative

Multiplication by a __negative__ number implies a __change__ in direction.   Here is an example.

> The climate of the Earth seems to be getting
> warmer.   Some glaciers (sheets of ice) are
> losing as much as 50 cm in length a year.
> (Call the rate -50 cm per year.)
>
> If the rate continues, what will be the length
> of the glacier two years from now?   (Call
> the time 2.)
>
> > Answer: Clearly, the glacier will be 100 cm
> > shorter then it is now.   We say
> > that $-50 \cdot 2 = -100$
>
> Suppose this rate has been the same for years.
> How long was the glacier 30 years ago?
> (Time = -30.)
>
> > Answer: 1500 cm longer than it is now.
> > We say that $-50 \cdot -30 = 1500$.

In other words, if we have a negative rate (loss) as years go on,

then years ago (which is also negative) there was more than there

is now.   The two negatives combine to give a positive result.

multiplier	direction	direction
-50/ yr	• years from now =	loss
-50/ yr	• years ago =	gain

In general:
Negative multipliers
switch direction.

multiplier	direction	direction
negative •	positive	= negative
negative •	negative	= positive

Examples: Multiplication by Negative Numbers

1. $-4 \cdot -9 = 36$ That is, multiplying -9 by -4 changes the direction of -9.

2. $-\frac{1}{2} \cdot 14 = -7$ Multiplying 14 by $-\frac{1}{2}$ changes the direction of 14.

3. $-1 \cdot 429 = -429$

4. $-6 \cdot -\frac{1}{3} = 2$

That multiplication reverses direction is seen by the size change of $-\frac{2}{3}$ pictured below.   The longer figure is the original.

Up is turned into down, right to left (and vice versa).

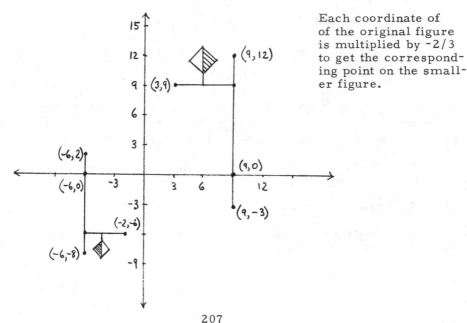

Each coordinate of of the original figure is multiplied by $-2/3$ to get the corresponding point on the smaller figure.

207

In this way, the size change model applies to negative as well as positive numbers. It is the most general of all models for multiplication.

Size Change Model
for
Multiplication:

> Let x be a scale factor. Then xy is
> x times as far from the origin as y is,
> and xy and y are:
> in the same direction if x is positive,
> in opposite directions if x is negative.

Questions covering the reading

1-4. You are an employer. Money into the business is positive. Money out is negative. What number represents each quantity?

1. pay of $2.50 an hour   2. pay of $8.00 an hour

3. time-and-a-half if normal pay is $7 an hour

4. time-and-a-half if normal pay is $4.37 an hour

5-12. Multiply:

5. $4 \cdot -6$ 6. $\frac{2}{3} \cdot -5$ 7. $\frac{3}{2} \cdot -\frac{8}{9}$ 8. $40 \cdot -11$

9. $3 \cdot -1$ 10. $1 \cdot -\frac{1}{2}$ 11. $1.4 \cdot -1.4$ 12. $\frac{2}{3} \cdot -\frac{1}{8}$

13-17. Refer to the glacier example in the lesson.

13. If the rate continues, what will be the length of the glacier 10 years from now?

14. What multiplication problem can get the answer to Question 13?

15. Suppose the rate has been the same for years. What was the length of the glacier 25 years ago?

16. What multiplication problem can get the answer to Question 15?

17. Review. Is 50 cm shorter or longer than the length of this page?

18-25. Multiply:

18. $-4 \cdot -3$ 19. $-10 \cdot -10$ 20. $-\frac{5}{2} \cdot -4$ 21. $-\frac{2}{3} \cdot -\frac{3}{2}$

22. $-1 \cdot -7$ 23. $-1 \cdot -1$ 24. $-5 \cdot -\frac{1}{5}$ 25. $-11 \cdot -6$

26. Give the size change model for multiplication.

27. If x is positive and y is negative, then xy is _____.

28. If x is negative and y is negative, then xy is _____.

29. If x is negative and y is positive, then xy is _____.

30-33. Refer to the drawing on p. 207. What point on the smaller figure corresponds to the given point on the original figure?

30. (9, 12)   31. (9, -3)   32. (3, 9)   33. (6, 13)

34. Copy this drawing and apply a size change of -2 to the figure. Put the resulting figure on the same graph. Be as accurate as you can.

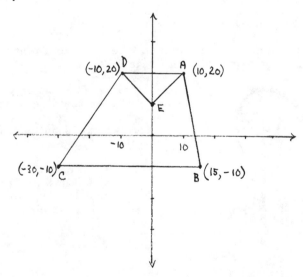

Questions testing understanding of the reading

1-4. Find the products. Your answers should form a pattern.

1. $3 \cdot 2; \ 3 \cdot 1; \ 3 \cdot 0; \ 3 \cdot -1; \ 3 \cdot -2; \ 3 \cdot -3; \ 3 \cdot -4$

2. $-\dfrac{3}{2} \cdot 4; \ -\dfrac{2}{2} \cdot 4; \ -\dfrac{1}{2} \cdot 4; \ 0 \cdot 4; \ \dfrac{1}{2} \cdot 4; \ \dfrac{2}{2} \cdot 4$

3. $-10 \cdot -8; \ -5 \cdot -8; \ 0 \cdot -8; \ 5 \cdot -8; \ 10 \cdot -8; \ 15 \cdot -8$

4. $-\dfrac{1}{3} \cdot 8; \ -\dfrac{1}{3} \cdot 5; \ -\dfrac{1}{3} \cdot 2; \ -\dfrac{1}{3} \cdot -1; \ -\dfrac{1}{3} \cdot -4; \ -\dfrac{1}{3} \cdot -7$

5-7. Call clockwise a negative direction, counterclockwise positive. (This is custom.) Gear G has 30 teeth, gear H has 15 teeth.

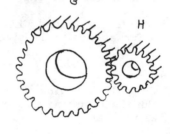

5. What will happen to gear H if:
   (a) G is turned 8 revolutions clockwise.
   (b) G is turned 2 revolutions counterclockwise.
   (c) What multiplication problems could help answer parts (a) and (b)?

6. (a) If H is turned 12 revolutions counterclockwise, what will happen to G?
   (b) What multiplication problem could help answer part (a)?

7. Let x represent turns of gear G. Let y represent turns of gear H. How are x and y related?

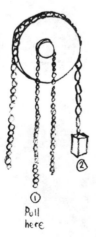

Pull
here

8-10. The larger pulley is 3 times the size of the smaller pulley. ① and ② are chains. A weight is on the end of chain 2.

8. If the end of chain 1 is pulled down 50 cm, what happens to the weight on chain 2?

210

9.   If the end of chain 1 goes up 25 cm, what happens to the
      weight on chain 2?

10.   Let x represent movement of the end of chain 1. Let
      y represent movement of the end of chain 2. How are
      x and y related?

11-14. Call clockwise negative, counterclockwise positive.
Pulley C has 2 times the circumference of pulley D.

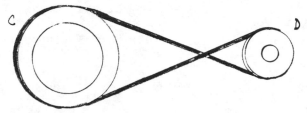

11.   If pulley C is turned -1/2 revolutions, how is pulley D turned?

12.   If pulley D is turned 100 revolutions, how is pulley C turned?

13.   If pulley C is turned r_C revolutions and moves pulley D r_D
      revolutions, how are r_C and r_D related?

14.   How could you get both pulleys to move clockwise at the same
      time?

15-17. Fahrenheit (U.S.) and Celsius (everywhere else) temperatures are related by the formula

$$F = \frac{9}{5} \cdot C + 32.$$

15.   Would -5° Celsius be good weather for skiing?

16.   At one of these numbers, the Fahrenheit and Celsius temper-
      atures are equal. Which number is it?

      (a) -10°   (b) -20°   (c) -30°   (d) -40°

17.   The lowest temperature possible ("Absolute zero") is -273.16°
      Celsius. What is the corresponding Fahrenheit temperature?

18-29. Simplify:

18. $10 \cdot -5 \cdot -1 \cdot -2$

19. $-1 \cdot -2 \cdot -3 \cdot -4$

20. $-\frac{2}{3} \cdot 6 \cdot -3$

21. $8 \cdot -2 \cdot \frac{1}{3}$

22. $-3 \cdot -4 \cdot x$

23. $6 \cdot -\frac{1}{12} \cdot y$

24. $-2 \cdot 2 \cdot -2 \cdot 2 \cdot -2$

25. $-1 \cdot -1 \cdot -1 \cdot -1 \cdot -1 \cdot -1$

26. $3 \cdot -4 + 6 \cdot -2$

27. $9 - -3 \cdot -4$

28. $-1 - -2 \cdot 5$

29. $-100 \cdot -2 \cdot -3 + 500 \cdot -4$

Lesson 7

Multiplication by 1 and -1

When you receive 100% of what you wanted, you get exactly
what you asked for.   That is:

$$100\% \cdot \text{amount} = \text{amount}$$

More simply, 100% = 1. So for any real number x,

$$\underline{1 \cdot x = x}.$$

The number 1 is called the <u>multiplicative identity</u>. Multiplying
by 1 keeps a number's identity. If you apply a scale factor of 1
to a figure, the result is a figure of the same size. The scale
1:1 is often called "actual size" for this reason.

A scale factor of ⁻1 also results in a figure of the same size as the original. But the directions are reversed. This is shown by the drawing. The original figure is at the bottom. All coordinates have been multiplied by ⁻1 to get the dotted figure.

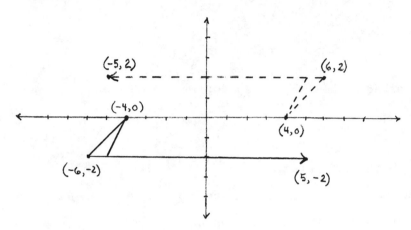

Reversing direction is the same as taking the opposite of a number. So for any real number x,

$$\underline{^-1 \cdot x = {^-}x.}$$

Although this is an important property of ⁻1, it is usually not given a name. We call it the <u>oppositing property of ⁻1.</u>

Examples: Applying the oppositing property.

1. ⁻a • ⁻b • ⁻c = ⁻1 • a • ⁻1 • b • ⁻1 • c   (Oppositing property)

            = ⁻1 • ⁻1 • ⁻1 • a • b • c   (Assemblage property)

            = ⁻1 • abc   (⁻1 • ⁻1 = 1; Identity)

            = ⁻abc   (Oppositing property)

213

2. $6 + {}^-1 \cdot x = 6 + {}^-x$ (Oppositing property)

                $= 6 - x$ (Def. of subtraction)

From the oppositing property, it is easy to see that

$$^-a \cdot {}^-b = ab$$

$$^-a \cdot b = {}^-ab = a \cdot {}^-b$$

You are asked to show why these are true in the questions which follow.

Questions covering the reading

1.   Multiplication by _____ keeps a number the same.

2.   Multiplication by _____ changes a number to its opposite.

3.   What is the oppositing property of $^-1$?

4.   What number is the multiplicative identity?

5-12. Simplify:

5. $^-1 \cdot x$ 6. $a + {}^-1 \cdot b$

7. $^-1 \cdot {}^-d$ 8. $^-5 \cdot {}^-e$

9. $^-x \cdot {}^-y$ 10. $^-7 \cdot {}^-m$

11. $^-4a \cdot {}^-1$ 12. $^-w \cdot x \cdot {}^-y \cdot z$

13.   What happens when a scale factor of 1 is applied to a figure?

14.   What happens when a scale factor of $^-1$ is applied to a figure?

214

15. Copy the graph at right.
Multiply all of the coor-
dinates of the points by
-1. Put the new points
on the same graph and
connect them.

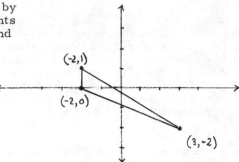

(-2,1)

(-2,0)

(3,-2)

Questions testing understanding of the reading

1-4. Simplify:

1. $^-x \cdot ^-y + xy$

2. $3 + ^-1 \cdot a + ^-1 \cdot b$

3. $d - ^-1 \cdot e - ^-1 \cdot f$

4. $^-m \cdot ^-n \cdot ^-p + g + mnp$

5. If $^-x = mx$, then m = _____.

6. If $y = ny$, then n = _____.

7. Give the property for each step. The steps show that
$^-x \cdot ^-y = xy$.

   (a) $^-x \cdot ^-y = ^-1 \cdot x \cdot ^-1 \cdot y$
   (b) $\qquad = ^-1 \cdot ^-1 \cdot x \cdot y$
   (c) $\qquad = 1 \cdot xy$
   (d) $\qquad = xy$

8. Give the property for each step. The steps show that
$x \cdot ^-y = ^-xy$.

   (a) $x \cdot ^-y = x \cdot ^-1 \cdot y$
   (b) $\qquad = ^-1 \cdot x \cdot y$
   (c) $\qquad = ^-xy$

9. Give the steps and properties which show that $^-x \cdot y = ^-xy$.

215

Chapter Summary

Lesson 1: The Repeated Addition Model for Multiplication
2: The Ordered Pair Model for Multiplication
3: The Area Model for Multiplication
4: The Assemblage Properties of Multiplication
5: The Size Change Model for Multiplication
6: Multiplication with Negative Numbers
7: Multiplication by 1 and $^-$1

There are four basic models for multiplication.

Model	Numbers Involved
repeated addition	multiplying by a positive integer
ordered pair	nonnegative integers
area	positive real numbers
size change	all real numbers

The models help in understanding some of the properties of multiplication. The area model is particularly helpful in showing the commutative, associative, and assemblage properties.

The size change model helps in showing multiplication by negative numbers. Multiplication by a negative number implies a change in direction. The numbers 1 and $^-$1 have special properties. 1 is the multiplicative identity: For any real number x, $1 \cdot x = x$. $^-$1 possesses the opposing property: $^-1 \cdot x = ^-x$. From these properties, it is easy to show that

$$^-a \cdot {}^-b = ab$$

$$a \cdot {}^-b = {}^-ab = {}^-a \cdot b$$

216

CHAPTER 5

MODELS FOR DIVISION

Lesson 1

The Splitting-Up Model for Division

In this short chapter, you will study the basic uses of division. Division is related to multiplication just like subtraction is related to addition. So for every model for multiplication, there is a corresponding model for division.

The most basic model arises from situations like the next one.

Question: Let 430 objects be split up evenly among 5 people. How many does each person get?

Answer: Divide 430 by 5. $\frac{430}{5} = 86$.

So each person gets 86 objects.

The above question is an instance of the splitting-up model of division.

Splitting-Up model for division:	If x objects are split up evenly into y compartments, there are $\frac{x}{y}$ objects in each compartment.

Suppose you split a length into 6 parts. You can get the length of each part by taking $\frac{1}{6}$ of the original length.

In this way, a division problem (splitting up) is converted into a multiplication problem (size change).

$$\begin{array}{ccc} \text{split x} \\ \text{into 6 parts} \end{array} = \begin{array}{c} \text{change x by the} \\ \text{scale factor 1/6} \end{array}$$

$$x \div 6 = x \cdot \frac{1}{6}$$

In general, division is defined so that all divisions can be converted to multiplications.

Definition of
   division:

$$x \div y = x \cdot \frac{1}{y}$$

The number $\frac{1}{y}$ is called the <u>reciprocal</u> of y. A number and its reciprocal multiply to 1. Here are some numbers and their reciprocals.

number y	2	$\frac{3}{5}$	$-\frac{3}{5}$	1.25	$\frac{a}{b}$
reciprocal $\frac{1}{y}$	$\frac{1}{2}$ or .5	$\frac{5}{3}$	$-\frac{5}{3}$	.8	$\frac{b}{a}$

The definition of division says that instead of dividing by y, you can multiply by the reciprocal of y. You probably learned how to divide fractions in this way.

divide
by $\frac{2}{5}$
$$\frac{4}{3} \div \frac{2}{5} = \frac{4}{3} \cdot \frac{5}{2}$$
$$= \frac{10}{3}$$
multiply by
reciprocal of $\frac{2}{5}$

The definition can also be used to simplify some fractions.

$$\frac{-8}{4} = -8 \cdot \frac{1}{4} = -2$$

You already did some division with negative numbers in calculating means.

Questions covering the reading

1. 1400 tickets to graduation are to be split up among 200 graduates. How many tickets will each graduate get?

2. 1400 tickets to graduation are to be split up among 203 graduates. How many tickets will each graduate get?

3. Suppose 100 tickets are to be split up among g graduates. How many tickets will each graduate get?

4. What is the splitting-up model for division?

5. A candy bar is split into 3 equal lengths. You can find the length of one of the little parts by multiplying the original length by _____.

6. A job requiring 50 hours to do is split up among p people. Assume each person will work the same amount of time. You can find out how long each person will have to work by dividing 50 by _____ or by multiplying 50 by _____.

7. What is the definition of division?

8-9. How can the definition of division be applied to do each problem?

8. $\dfrac{9}{11} \div \dfrac{8}{7} =$ 9. $\dfrac{a}{b} \div 5 =$

10. Simplify $\dfrac{-25}{10}$.   11. Simplify $\dfrac{-21}{3}$.

12-19. Give the reciprocal of each number.

12. 11   13. -5   14. -2   15. $\dfrac{1}{3}$

16. $\dfrac{2}{5}$ 17. $\dfrac{9}{4}$ 18. 1.8   19. .0625

20. Choose all correct answers. The reciprocal of 1.25 is

(a) .75   (b) $\dfrac{4}{5}$ (c) .8   (d) $\dfrac{3}{4}$

21. Choose all correct answers. The reciprocal of $\dfrac{3}{5}$ is

(a) $\dfrac{5}{3}$ (b) $-\dfrac{5}{3}$ (c) 1.6   (d) $1.\overline{6}$

Questions testing understanding of the reading

1. True or False: If you divide 5834 by 8, you will get the same answer as if you multiply 5834 by .125.

2. True or False: If a is the reciprocal of b, then b is the reciprocal of a.

3. Pick all correct answers. Suppose x and y are reciprocals. Then:

(a) $x = \dfrac{1}{y}$ (b) $xy = 1$ (c) $y = \dfrac{1}{x}$ (d) $\dfrac{x}{y} = 1$

4. Simplify: $x \cdot y \cdot \dfrac{1}{y}$ 5. Simplify: $\dfrac{1}{z} \cdot 5 \cdot x \cdot z$

6-11. Apply the definition of division to simplify.

6. $\dfrac{3}{5} \div 4$ 7. $4 \div \dfrac{3}{5}$

8. $115 \div \dfrac{a}{b}$ 9. $\dfrac{x}{y} \div 6$

10. $\dfrac{a}{b} \div \dfrac{c}{d}$ 11. $75 \div \dfrac{x}{4}$

12-17.   These problems look different from Questions 6-11 but they aren't.

12. $\dfrac{\frac{9}{4}}{3}$

13. $\dfrac{10}{\frac{2}{5}}$

14. $\dfrac{120}{\frac{x}{y}}$

15. $\dfrac{\frac{n}{2n}}{5}$

16. $\dfrac{\frac{e}{f}}{\frac{g}{h}}$

17. $\dfrac{2}{\frac{1}{x}}$

18.   How many $\frac{3"}{8}$ thick strips can be made from a piece of wood 12" thick? Show how this problem can be considered as a division problem.

19.   How many quarters are in $20.50? How can this problem be done as a division problem?

Questions for discussion

1.   Explain how finding the mean can be considered as applying the splitting-up model of division.

2.   Refer to Question #2, p. 219. How would you split up the tickets?

Lesson 2

Probability of Outcomes

Imagine tossing a perfectly balanced 6-sided die. You would expect each side to come up about $\frac{1}{6}$ of the time. The number $\frac{1}{6}$ is the <u>probability</u> of tossing a 2. It is also the probability of tossing a 3, and so on.

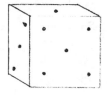

Definition:

> If an experiment has n equally likely outcomes, then the <u>probability</u> of each outcome is $\frac{1}{n}$.

For example, in the experiment "flipping a coin," there are two outcomes. If you assume that these outcomes are equally likely, then the probability of heads is $\frac{1}{2}$ and the probability of tails is $\frac{1}{2}$.

When each outcome of a coin (or die or other such thing) has the same probability, we call the coin <u>fair</u> or <u>unbiased</u>. In practice, fair coins do not occur. (Even if a fair coin were around, we could not tell. Flipping 1000 times and getting 500 heads could happen even with a weighted coin.) We can only <u>imagine</u> flipping a fair coin and ask what would happen.

When each outcome of an experiment is assumed to have the same probability, we say that the outcomes occur <u>at random</u> or <u>randomly</u>. Randomness is also something which can only be imagined. Many experiments and games are designed in hopes that the outcomes will occur in as close to random fashion as possible.

When you actually have a coin, you can only guess whether it is fair. To test the fairness, you might flip the coin 1000 times. Suppose 510 heads occurred. The relative frequency of heads is then $\frac{510}{1000}$. This is close enough to $\frac{1}{2}$ to make one think the coin is balanced. In the long run (say after 100,000 tosses), you would want the relative frequency to be even closer to $\frac{1}{2}$ before you would believe the coin is balanced. But you can never know for sure.

Notice the differences between a relative frequency and a probability. A relative frequency always results from an actual experiment. A probability is theorized--it is theoretical or assumed, often from the relative frequency you would expect in the long run. We often use actual experiments and relative frequencies to test whether probabilities have been selected in a reasonable way.

But probabilities have much in common with relative frequencies. The largest a probability can be is 1; this occurs if the event must happen. The smallest a probability can be is 0; this occurs if the event is impossible. (For example, the probability of tossing a 7 with a single die is 0.) All probabilities are between 0 and 1, inclusive. And the more likely an event is thought to be, the closer its probability is to 1.

Examples: Probabilities

1. Suppose you expect an event to occur 90% of the time. Then you might say its probability of occurring is $\frac{9}{10}$ (or .9). You could even use the percentage 90%, as TV weathermen do with precipitation probabilities.

2. In the past, about 1 in 86 births has resulted in twins. You could choose $\frac{1}{86}$ as the probability of twins. You would then be using a relative frequency to estimate the probability of twins. This is not a bad idea because there does not seem to be any other way of calculating the probability.

What probabilities do is split up the number 1 (or 100%, if you like). The splitting is similar to that done in a circle graph. And if you split 1 into n parts of equal size, each part is $\frac{1}{n}$. So the basic idea behind probability is the splitting-up model of division.

Questions covering the reading

1. What is a fair coin?   2. What is an unbiased die?

3. Do unbiased coins exist?   4. Do fair dice exist?

5. For a fair coin, what is the probability of tails?

6. For a fair die, what is the probability of throwing a 4?

7. For a fair die, what is the probability of throwing an 8?

8. Suppose a situation has 400 equally likely outcomes. What is the probability of each outcome?

9. Repeat Question 8 for a situation with n outcomes.

10. When do outcomes occur randomly?

11. Review. Define: relative frequency.

12. Name one difference between relative frequency and probability.

13. Name two similarities between relative frequency and probability.

1. Suppose you toss a coin 4 times and the coin lands heads up twice. Is the coin a fair coin?

2. Imagine tossing a fair coin 100 times. Will 50 heads come up?

3-6. Give the probability which would normally be used in each situation.

3. You expect an event to occur 2 out of 3 times.

4. You expect an event to occur 5 times as often as it doesn't occur.

5. You expect an event to occur 9 times as often as it doesn't occur.

6. You expect an event to occur $\frac{3}{4}$ of the time.

7. A card is randomly picked from a deck of 52 cards. What is the probability that the card is the ace of spades?

8. At a carnival, you tell a "mind-reader" the year you were born. The mind-reader tries to guess your birthday. If the mind-reader guesses randomly, what will be the probability of guessing correctly?

9-11. If you guess randomly, what is the probability of being correct on a:

9. true-false question.   10. always-sometimes-never question.

11. multiple choice question with c choices.

Lesson 3

Probability of Events

Recall that an experiment is a situation being studied. The possible results of an experiment are outcomes. You have covered all outcomes to an experiment if (1) no two outcomes occur at the same time and (2) no possible outcome is missing.

Example 1:   An experiment with cards: You pick a card from a normal deck of 52 cards.

This experiment has 52 outcomes. One outcome is "picking the 9 of hearts," written "9H" for short.

In this example, "picking a king" is not an outcome. It involves 4 possible outcomes: KS (king of spades), KH, KC, and KD. We say that "picking a king" is an _event_. "Picking a king" = $\{$KS, KH, KC, KD$\}$.

Definition:

An _event_ is a set of outcomes.

The event "picking a diamond" contains 13 outcomes. The event "picking a black queen" has 2 outcomes. The event "picking a card" has 52 outcomes.

An event occurs whenever any of its outcomes occur. So it is natural to make the following definition for calculating the probability of an event.

Definition:

> The probability of an event is the sum of the probabilities of its outcomes.

Example 2: Suppose a card is randomly picked from a deck. What is the probability of the event "picking a king"?

Solution: Since the picking is at random, each outcome has probability $1/52$.

$$\text{picking a king} = \{KS, KH, KD, KG\}$$

$$\begin{pmatrix}\text{the probability} \\ \text{of picking a king}\end{pmatrix} = \begin{pmatrix}\text{the sum of the probabilities} \\ \text{of the four outcomes}\end{pmatrix}$$

$$= 1/52 + 1/52 + 1/52 + 1/52$$

$$= 4/52$$

In the situation of Example 2, there are 4 kings and 52 cards. The answer suggests that there is an easy pattern for probabilities when outcomes occur randomly. And there is, by the same type of argument used in Example 2.

Probability
of an Event
Theorem:

> Assume outcomes in an experiment occur randomly. Then the probability of an event
>
> $$= \frac{\text{number of outcomes in the event}}{\text{number of outcomes in the experiment}}$$

227

Example 3:   A number is randomly selected from $\{1, 2, 3, 4, 5\}$.
                  What is the probability of selecting an even number?

Answer: There are 5 outcomes possible. The event "sel-
                  ecting an even number" contains the outcomes 2
                  and 4. So the probability is $\frac{2}{5}$.

The next examples involve very common and important situa-

tions. The basic problem in each is to count how many outcomes

are possible. The ordered pair model of multiplication--perhaps

extended to more than 2 blanks--is of help.

Example 4:   Suppose <u>two</u> dice are thrown. If outcomes occur
                  randomly, what is the probability of getting a
                  sum of 7?

Solution: Each die has 6 possibilities. So there are 6·6
                  possible outcomes. They are listed here. The
                  outcome 2 1 means a 2 was thrown on the first
                  die, a 1 on the second die.

1 1	2 1	3 1	4 1	5 1	(6 1)
1 2	2 2	3 2	4 2	(5 2)	6 2
1 3	2 3	3 3	(4 3)	5 3	6 3
1 4	2 4	(3 4)	4 4	5 4	6 4
1 5	(2 5)	3 5	4 5	5 5	6 5
(1 6)	2 6	3 6	4 6	5 6	6 6

The event "getting a 7" consists of the circled outcomes.

Since the outcomes occur at random:

probability of getting a 7 $= \dfrac{\text{no. of outcomes in event}}{\text{no. of outcomes in experiment}} = \dfrac{6}{36} = \dfrac{1}{6}$

228

Example 5:   Some race tracks offer a "quinella," a prize for
              correctly predicting the winners of 5 straight
              races. Suppose the following numbers of horses
              run in the 5 races.

                        Race 1:   8 horses
                             2:      10 horses
                             3:      7 horses
                             4:      6 horses
                             5:      11 horses

              If you bet at random, what is your probability
              of winning?

Solution: You need to know how many betting combina-
              tions there are. Think of filling in five blanks
              with the number of horses in each race.

              _____   _____   _____   _____   _____
              Race 1   2   3   4   5

              Applying the ordered pair model for multiplica-
              tion, the number of betting combinations is the
              product of 8, 10, 7, 6, and 11. So the proba-
              bility of picking the winning betting combination
              is $\frac{1}{36960}$, a very small number.

Questions covering the reading

1.   Define: event.

2.   Define: probability of an event.

3-8. A card is selected from a deck of 52 cards. (a) How many
outcomes are in each event? (b) If the selection is random, what
is the probability of the event?

3.   picking a queen   4. choosing a club

5.   picking a 10   6. selecting a red card

7.   choosing the 2 or 3 of hearts

8.   choosing a 17 of spades

229

9. Suppose 10 horses run in the 1st race and 8 in the 2nd race. You win the "daily double" if you correctly select the winner in each race. If you guess randomly, what is your probability of winning the daily double?

10. Repeat Question 9 if 5 horses run in each race.

11. When 3 multiple-choice questions have 5 choices each and only 1 is correct, what is the probability of randomly guessing all 3 correctly?

12-23. Suppose 2 fair dice are thrown. Calculate the probability of getting a sum of:

12. 2	13. 3	14. 4	15. 5
16. 6	17. 7	18. 8	19. 9
20. 10	21. 11	22. 12	23. 13

Questions testing understanding of the reading

1-5. A large department store needs 6 stock people, 10 salespersons, and 5 cashiers for the summer. Suppose you apply for and get a job at this store. Think of this situation as having 21 outcomes.

1. How many outcomes make up the event "not working as a stock person?"

2. How many outcomes make up the event "working at the store?"

3. If people are randomly assigned to jobs, what is the probability you will be a salesperson?

4. Knowing no other information, what is the probability that you will work as a cashier?

5. Knowing no other information, what is the probability that you will not work as a cashier?

6-9. In hockey, a team can win, lose, or tie. Suppose, on the basis of past experience, you feel that in a big game:

> the probability your team wins is .4.
> the probability your team loses is .5.
> the probability of a tie is .1.

Give the probability for the event.

6.   Your team does not lose.   7. Your team does not win.

8.   Your team plays.   9. Your team does not play.

10-15. A number n is randomly selected from the set {1, 2, 3, 4, 5, 6, 7, 8, 9, 10} . Give the probability of the event.

10.   n is even.   11. n is divisible by 3.

12.   n is prime.   13. $n \cdot n > 50$

14. $6n \leq 20$ 15. $\dfrac{9}{n} = 1$

16-17. Suppose you randomly guess at a multiple-choice question with 5 choices: A, B, C, D, E. If D is the correct choice, name the outcomes in each event and the probability of that event.

16.   getting the correct answer

17.   getting a wrong answer

18.   A restaurant serves 6 different appetizers, 10 different main courses, and 9 different desserts. If two people would randomly select their meal (one of each), what is the probability that they would select the same meal?

19.   Blank A can be filled with any of the numerals 1, 2, or 3. Blank B can be filled with 4 or 5, blank C with 6, 7, or 8.

<u>　　　</u>　　　<u>　　　</u>　　　<u>　　　</u>
　　　　A　　　　　　　B　　　　　　C
(a) How many different 3-digit numerals can be formed in this way?
(b) Write down <u>all</u> of them.
(c) Randomly pick one of the answers you wrote in part (b). What is the probability that you will pick the same one that your teacher picks?

20-21. Refer to Example 5 on page 229.

20. Suppose 12 horses run in the 4th race instead of 6. How does this affect the chances of winning?

21. Suppose only 2 horses run in each race. (a) How many betting combinations would then be possible? (b) If selections are made at random, what is your probability of winning?

Lesson 4 (Optional)

Calculating Probabilities Where Counting Is Impossible

In Lessons 2 and 3, you have calculated probabilities by counting outcomes. It is not always possible or advisable to do this. Here is a simple problem in which counting is impossible.

(1) The circle at right represents a disc with a spinner. Suppose that any position of the spinner is equally likely. When the spinner is spun, what is the probability that it lands in the region A? (Think about this one. The answer is given on the next page.)

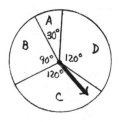

(2) A dot moves back and forth at a uniform speed from 0 to 1 to 0 to 1, and so on. If a picture is taken of the dot, what is the probability that the dot will be between $\frac{1}{2}$ and $\frac{2}{3}$? (Answer on page 233. Guess at an answer first.)

232

(3)   A person throws darts at the
square diagram at right. What is
the approximate probability of
landing in a dark area?

Problem (3) is more difficult than (1) or (2), but it is easy

to understand what is going on from this problem. We ask: What

percentage of the diagram is dark? That percentage approximates

the probability. For this example, the percentage is

$$\frac{\text{area of dark region}}{\text{area of square}} \ .$$

In general, here is the definition of these kinds of probabil-

ities (given equally likely potential of being anywhere in the set).

Probability of landing in subset	=	measure (area, length, etc.) of subset
		measure of entire set

For example, in problem (1),
think of the set as being the circle
itself. The probability that the
spinner lands in region A is

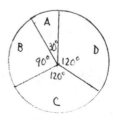

$$\frac{30^{\circ}}{360^{\circ}} \text{ or } \frac{1}{12} \ .$$

233

In problem (2), the probability that the dot will be between $\frac{1}{2}$ and $\frac{2}{3}$ is

$$\frac{\text{length of AB}}{\text{length of XY}} = \frac{\frac{2}{3} - \frac{1}{2}}{1} = \frac{1}{6}$$

Counting is a special type of measuring. So the Probability of an Event Theorem given on page 227 is a special case of the more general definition given in this lesson.

Questions covering the reading

1-5. In the spinner situation pictured on p. 232, what is the probability of landing in region

1. A.　　2. B.　　3. C.　　4. D.　　5. A or D.

6.　　When counting is not possible, how are probabilities often defined?

7-10.　A dot moves at a uniform speed back and forth between 0 and 10 on the number line. If a picture is taken of the dot, what is the probability that the dot will be between

7.　　0 and $\frac{1}{2}$.　　8. 1 and 4.　　9. 0 and 10.　　10. 9 and 9.01.

Questions testing understanding of the reading

1-4.　A clock with a second hand is stopped by a power failure. What is the probability that the second hand lies between

1.　　12 and 3　　2. 3 and 7　　3. 5 and 6　　4. 11 and 1

5-8. Here is another situation where counting is difficult. In the U.S., 1970 population was distributed by age as follows:

under 14	27. 8%
14-24	18. 6%
25-44	23. 5%
45-64	20. 7%
65 and over	9. 5%

Choose a person in the U.S. in 1970 at random. What is the probability that the person is:

5. under 14.

6. over 64.

7. under 25.

8. between 25 and 64.

C   9-10.   The land area of the Earth is about 57,510,000 sq miles. The water surface area is 139,440,000 sq miles. What is the probability that a meteor hitting the surface of the Earth will:

9. fall on land.

10. fall in water.

Lesson 5

The Scale Comparison Model for Division

A probability measures how often you expect something to happen compared (by division) to how often it could happen. In this way, numbers used in probability are numbers being used in comparison. As you learned in Chapter 1, comparison by division is very common. It leads to ratios and rates. Here is a typical example.

235

In 1960 the population of the U.S. was about 179 million. In 1970 the population was about 204 million. Now divide.

$$\frac{204 \text{ million}}{179 \text{ million}} \approx 1.20$$

The quotient 1.20 indicates:

(1) There were about 1.20 times as many people in the U.S. in 1970 as in 1960.

(2) The U.S. population in 1970 was 120% of the U.S. population in 1960. (This is because 1.20 = 120%)

(3) There was a 20% increase in the U.S. population from 1960 to 1970.

The 1.20 is a scale factor. This type of comparison of numbers is called scale comparison. The name "scale" comes from its use in maps and other scale drawings.

Here are two drawings. The original A is at left. An enlarged picture B is at right.

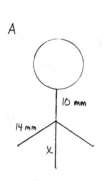

A

10 mm

14 mm

x

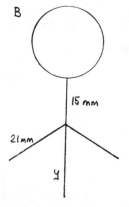

B

15 mm

21mm

y

Now divide any length in figure B by the corresponding length

in the original figure A. In every case you will get 1.5.

$$\frac{\text{length in enlarged figure}}{\text{corresponding length in original}} = \frac{15 \text{ mm}}{10 \text{ mm}} = \frac{21 \text{ mm}}{14 \text{ mm}} = 1.5$$

The lengths x and y marked on the figures also have this rela-

tionship. $\frac{y}{x} = 1.5$.

Scale comparison model for division:	$\frac{y}{x}$ is the scale of a size change which takes an original length x into a new length y.

Scale comparisons are found in blueprints, model airplanes

and other scale models, binoculars, telescopes, scale drawings,

etc. In each of these situations, the scale can be calculated by

division.

Example 1: Suppose an insect 1" long is detailed in a drawing
which is 4" long. What is the scale of the drawing?

Solution: Scale of drawing = $\frac{\text{length in drawing}}{\text{length in original}}$

$= \frac{4"}{1"} = 4$

The drawing is, as you could guess, 4 times
actual size. We could say that its scale is 4:1.

Look again at the drawings on page 236. The scale of 1.5

indicates that any length in figure B is 1.5 times the corresponding

length in figure A. This explains why the quotient is always 1.5.
When you divide:

$$\frac{\text{length in figure B}}{\text{length in figure A}} = \frac{1.5x}{x} = (1.5x) \cdot \frac{1}{x} \quad \text{(Definition of division)}$$

$$= 1.5 \cdot x \cdot \frac{1}{x} \quad \text{(Assemblage property of } \cdot \text{)}$$

$$= 1.5 \cdot 1 \quad \text{(Reciprocals multiply to 1)}$$

$$= 1.5 \quad \text{(1 is multiplicative identity)}$$

Here are four divisions which have been done in this lesson.

(1) $\frac{4''}{1''} = 4$ (2) $\frac{21\text{mm}}{14\text{mm}} = 1.5$

(3) $\frac{204 \text{ million}}{179 \text{ million}} \approx 1.20$ (4) $\frac{1.5x}{x} = 1.5$

In divisions (1) and (2) the units "cancel out." In division (3), you
can think of "million" as a unit. It also cancels out. That is,

$$\frac{204 \text{ million}}{179 \text{ million}} = \frac{204}{179} \approx 1.20$$

But you can think of million as being the number 1,000,000. Then
it is a factor common to the numerator and denominator. (A factor
is a number which is multiplied.) The factor can be removed by
the following process.

$$\frac{204 \cdot 1,000,000}{179 \cdot 1,000,000} = \frac{204}{179} \cdot \frac{1,000,000}{1,000,000} = \frac{204}{179} \cdot 1 = \frac{204}{179}$$

238

In division (4), x is a factor common to the numerator and denominator.

$$\frac{1.5x}{x} = \frac{1.5 \cdot x}{1 \cdot x} = 1.5$$

You have used this idea to simplify fractions.

Fraction
Simplification
Property:

> If the numerator and denominator of a fraction contain the same factor, that factor can be removed from each without changing the value of the fraction. In symbols,
>
> $$\frac{ax}{ay} = \frac{x}{y}$$

Examples: Simplifying Fractions.

1. $\dfrac{4y}{y} = 4$ 2. $\dfrac{15x}{15} = x$

3. $\dfrac{3x}{3y} = \dfrac{x}{y}$

4. $\dfrac{9a\,(b+2)}{3a} = \dfrac{3 \cdot (b+2) \cdot 3a}{3a} = 3\,(b+2)$ The common factor is 3a.

5. $\dfrac{95nb}{19nc} = \dfrac{19n \cdot 5b}{19n \cdot c} = \dfrac{5b}{c}$ The common factor is 19n.

The Fraction Simplification Property does not work if the number common to the numerator and denominator is added.

$$\frac{3 + x}{3 + y} \text{ cannot be simplified.}$$

Questions covering the reading

1. A letter .8 cm high is magnified by a magnifying glass to be 1.6 cm high. How powerful is the glass?

2. What is the scale on the drawing of a 6-foot man if the drawing is 8" high?

3-4. A photograph of a stone is known to be 5 times actual size.

3. If the stone is c cm long, then on the photograph it will be _____ cm long.

4. If you divide the length of the stone in the photograph by the actual length of the stone, what do you get?

5. What is the scale comparison model for division?

6-9. Drawing D is an enlargement of drawing C.

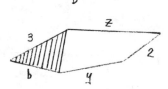

6. Name four fractions which are equal to $\frac{y}{x}$.

7. What is the scale factor?

8. Calculate b.   9. Calculate z.

10. What is the Fraction Simplification Property?

11-22. (a) Name the factor common to the numerator and denominator. (b) Simplify.

11. $\dfrac{8x}{2}$

12. $\dfrac{144a}{3}$

13. $\dfrac{10y}{y}$

14. $\dfrac{\frac{1}{2} \cdot B}{B}$

15. $\dfrac{3(m+5)}{m+5}$

16. $\dfrac{2(3t-1)}{3t-1}$

17. $\dfrac{14y}{2y}$

18. $\dfrac{6v}{11v}$

19. $\dfrac{49ab}{14b}$

20. $\dfrac{100xy}{2x}$

21. $\dfrac{50m}{50n}$

22. $\dfrac{2z}{150}$

Questions testing understanding of the reading

1 2. Evaluate each expression when $x = 40,000$ and $y = 200$.

1. $\dfrac{43x}{x}$

2. $\dfrac{9xy}{18xy}$

3-4. Evaluate each expression when $A = 20$, $B = 19$, and $C = 18$.

3. $\dfrac{ABC}{5BC}$

4. $\dfrac{430C}{215C}$

5. A student thought that $\dfrac{6+x}{3+x}$ should simplify to 2. (a) By substitution, show that the student was wrong. (b) Why doesn't the Fraction Simplification Property apply to this problem?

6. Show that $\dfrac{x+5}{y+5}$ is not always equal to $\dfrac{x}{y}$.

7-10. Compare the two given numbers by dividing. Then use the quotient in a sentence.

   Example: Fifi receives $5 a week allowance. Gigi gets
            $4 a week.
   Answer: 5/4 = 1 1/4. Fifi receives 1 1/4 times as much
            allowance as Gigi.

7. The U.S. population is about 213 million. Canada's population is about 22 million.

241

8. The length of that building is 80 meters. Its width is 50 meters.

9. A segment of length x on this drawing corresponds to a segment of length 3x on that drawing.

10. Her car gets 8 km per liter. Mine gets 10 km per liter.

11-12. Suppose the scale of a blueprint is 1:30.

11. An object with actual length x will be pictured having length _____ on the drawing.

12. If you divide the length of the object in the drawing by the actual length of the object, what will you get?

13. A map in an atlas lists its scale as 1:300,000. How is this scale calculated?

14. Give a reason for each step in this formal simplification of $\dfrac{kx}{x}$.

$$\text{(a)} \quad \frac{kx}{x} = (kx) \cdot \frac{1}{x}$$

$$\text{(b)} \quad = k \cdot \left(x \cdot \frac{1}{x}\right)$$

$$\text{(c)} \quad = k \cdot 1$$

$$\text{(d)} \quad = k$$

15-20. Simplify.

15. $\dfrac{405}{315}$

16. $\dfrac{axym}{xa}$

17. $\dfrac{2bqq}{3qb}$

18. $\dfrac{729}{1512}$

19. $\dfrac{4\ \mathbf{bat}}{\frac{1}{2}tab}$

20. $\dfrac{xy}{yy}$

Question for discussion

1. The fraction $\dfrac{8}{6}$ can be simplified to $\dfrac{4}{3}$.

In your earlier work, you may have heard this called "reducing" the fraction. The word "reducing" is not used in this book. Instead, "simplifying" is used. Why do you think this is done?

Lesson 6

The Rate Model for Division

Suppose you divide your height by your weight. Here is what happens if you are 5'8" (that is 68") tall and weigh 136 pounds.

$$\frac{68"}{136 \text{ lb.}}$$

Separate the units from the real numbers. Then use the word per to indicate the "division" of the units.

$$\frac{68 \text{ inches}}{136 \text{ pounds}} = \frac{68 \text{ inches}}{136 \text{ pounds}} = .5 \text{ inches per pound}$$

The answer is a rate. In this case, the rate says that $\frac{1}{2}$ inch in height is accounted for by a pound of weight.

You could divide the other way.

$$\frac{136 \text{ pounds}}{68 \text{ inches}} = \frac{136}{68} \frac{\text{lb}}{\text{in}} = 2\frac{\text{lb}}{\text{in}} = 2 \text{ pounds per inch}$$

There is a mean of 2 pounds per inch of height.

The most familiar rate is speed. Speed is calculated by dividing a distance by a time. For example,

$$\frac{250 \text{ km}}{3 \text{ hr}} = \frac{250}{3} \frac{\text{km}}{\text{hr}} = 83.\overline{3} \frac{\text{km}}{\text{hr}} = 83.\overline{3} \text{ km per hour}$$

This means that if one goes 250 km in 3 hr, then the mean rate is 83.$\overline{3}$ km per hour.

Rate model
for division:

$$\frac{x \text{ of unit } 1}{y \text{ of unit } 2} = \frac{x}{y} \text{ unit 1's per unit 2,}$$

the mean rate of unit 1's per unit 2.

Rates can be negative. Suppose you go on a diet and lose 2 kg in the next 12 days. You have lost $\frac{1}{6}$ kg per day. This can be calculated by doing a division problem with negative numbers.

$$\text{rate of weight gain} = \frac{\text{weight change}}{\text{time}} = \frac{-2 \text{ kg}}{12 \text{ days}} = \frac{-2}{12} \frac{\text{kg}}{\text{days}} = -\frac{1}{6} \text{ kg/day}$$

The same rate would happen if you looked back and noticed that 12 days ago you weighed 2 kg more. Then

$$\text{rate of weight gain} = \frac{\text{weight change}}{\text{time}} = \frac{2 \text{ kg more}}{12 \text{ days back}} = \frac{2}{-12} \frac{\text{kg}}{\text{days}} = -\frac{1}{6} \text{ kg/day}$$

This example shows that, as with multiplication, <u>division with a positive and a negative number yields a negative quotient</u>.

Specifically, $\frac{-2}{12} = \frac{2}{-12} = -\frac{2}{12}$

and, in general,

$$\frac{-x}{y} = \frac{x}{-y} = -\frac{x}{y}$$

Now let us consider a situation with division of negative numbers. Suppose you are working and earning $10 a day. You might have calculated your pay rate by noting that 4 days ago you had $40 less than you have now.

244

$$\text{rate} = \frac{\text{money gained}}{\text{time}} = \frac{-40 \text{ dollars}}{-4 \text{ days}} = \frac{-40 \text{ dollars}}{-4 \text{ days}} = \$10/\text{day}$$

In general, if something is increasing as you go forward, then it is decreasing as you go backward. In each case, the same positive rate would occur. Thus <u>the quotient of two negative numbers is positive</u>. Specifically,

$$\frac{-40}{-4} = \frac{40}{4} \quad \text{and in general,} \quad \boxed{\frac{-x}{-y} = \frac{x}{y}}$$

One or two examples are not enough to make a general statement. Mathematicians would not use rate to give them information about division of positive and negative numbers. They would probably use the definition of division and remember two things: the reciprocal of a positive number is positive. The reciprocal of a negative number is negative. Then:

$$\frac{\text{positive}}{\text{negative}} = \text{positive} \cdot \text{recip of negative} = \text{positive} \cdot \text{negative}$$

$$\frac{\text{negative}}{\text{positive}} = \text{negative} \cdot \text{recip of positive} = \text{negative} \cdot \text{positive}$$

$$\frac{\text{negative}}{\text{negative}} = \text{negative} \cdot \text{recip of negative} = \text{negative} \cdot \text{negative}$$

In each case the division problem is converted into a similar multiplication problem. <u>With respect to positive or negative directions, multiplication and division give the same answers</u>.

The simplification of rates looks very much like multiplication of fractions.

$$\frac{250 \text{ km}}{3 \text{ hrs}} = \frac{250}{3} \frac{\text{km}}{\text{hr}} \qquad \text{looks very much like} \qquad \boxed{\frac{ac}{bd} = \frac{a}{b} \cdot \frac{c}{d}}$$

You can use this idea to multiply fractions.

Examples: Multiplying fractions

1. $\quad \dfrac{-9}{4} \cdot \dfrac{2}{-3} = \dfrac{-18}{-12} = \dfrac{18}{12} = \dfrac{3}{2}$

2. $\quad \dfrac{2x}{3} \cdot \dfrac{-5}{9x} = \dfrac{-10x}{27x} = \dfrac{-10}{27} = -\dfrac{10}{27}$

3. $\quad \dfrac{8ab}{3a} \cdot \dfrac{4c}{6b} = \dfrac{32abc}{18ab} = \dfrac{32c}{18} = \dfrac{16c}{9}$

All of the simplifications of this lesson and the last lesson work because the numbers are multiplied or divided. When numbers are added, rates become difficult to calculate and fractions are difficult to simplify. (Recall that it is harder to add fractions than to multiply them.)

Questions covering the reading

1-4. Calculate your rate of weight gain or loss over time in each situation.

1.   15 days from now you weigh 3 pounds less.

2.   15 days ago you weighed 3 pounds more.

3.   10 days from now you weigh 1 kg more.

4.   10 days ago you weighed 1 kg less than you do now.

5-8. Calculate a rate related to the given information.

5. She is 171 cm tall and weighs 57 kg.

6. We travelled 400 miles in $9\frac{1}{2}$ hours.

7. In 3 days, I earned $40.

8. 6 cans of orange juice cost $2.00.

9-12. Perform the division and tell what your answer could mean.

9. $\dfrac{-7 \text{ lbs}}{2 \text{ wks}}$

10. $\dfrac{\$32.46}{9.2 \text{ hr}}$

11. $\dfrac{86 \text{ points}}{4 \text{ quarters}}$

12. What is "speed" and how is it calculated?

13. What is the rate model for division?

14-22. Simplify.

14. $\dfrac{-11}{-22}$

15. $\dfrac{-1400}{20}$

16. $\dfrac{800}{-400}$

17. $\dfrac{-20}{-50}$

18. $\dfrac{-112}{112}$

19. $\dfrac{6}{-14}$

20. $\dfrac{75}{-25}$

21. $\dfrac{-2x}{-x}$

22. $\dfrac{-b}{-a}$

23. What is the general pattern for multiplication of fractions?

24-32. Multiply. Then simplify.

24. $\dfrac{-1}{3} \cdot \dfrac{2}{-6}$

25. $\dfrac{-x}{y} \cdot \dfrac{-y}{x}$

26. $\dfrac{-2}{5} \cdot \dfrac{-5}{-4} \cdot \dfrac{-6}{-5}$

27. $\dfrac{200}{27} \cdot \dfrac{9}{400}$

28. $\dfrac{3}{x} \cdot \dfrac{x}{6}$

29. $\dfrac{A}{B} \cdot \dfrac{B}{A} \cdot \dfrac{A}{B}$

30. $\dfrac{1}{m} \cdot \dfrac{1}{n} \cdot mn$

31. $\dfrac{-2y}{7} \cdot \dfrac{14z}{-3y}$

32. $\dfrac{3x}{-4} \cdot \dfrac{4}{3}$

33-35. True or False. x/y is negative when:

33. both x and y are negative.

34. x is positive and y is negative.

35. x is negative and y is positive.

247

Questions testing understanding of the reading

1-2. Suppose $x = {}^-40$ and $y = 5$. True or False:

1. $\dfrac{{}^-x}{{}^-y} = \dfrac{x}{y}$

2. $\dfrac{{}^-x}{y} = \dfrac{x}{{}^-y}$

3-10. **Multiple choice.** Given that p is positive and n is negative. The choices are:
(a) is always positive
(b) is always negative
(c) can be either positive or negative

3. $p + n$　　　4. $p - n$　　　5. $n - p$　　　6. $-3pn$

7. pn　　　8. p/n　　　9. $2n/p$　　　10. $(p + n)(p - n)$

11. A car used up 20 gallons of gas in going 440 miles. One person said the rate of gas usage was 22 miles per gallon. Another said that the rate of gas usage was 1/22 gallon per mile. Who was right?

12. If you go x km and use 60 liters of gasoline, how many kilometers per liter are you getting in your driving?

13. A person gained w kilograms in t days on a diet. What was the rate of weight gain per day? Give a specific example to check your answer.

14-18. In each of these problems you are expected to multiply units as you might multiply fractions.

14. If a person drives for 6 hours at a mean rate of 70 kilometers per hour, how far has the person driven?

15. If a person buys 20 pencils at the rate of 2 pencils for 5 cents, how much has been spent?

16. A machine can sort 300 cards/minute. In 2 1/2 hours how many cards can be sorted?

17. Some material costs $2.25 per yard. How much will z yards of this material cost?

18. Suppose 500 people live on an average city block. Then on b blocks _____ .

19. Show that $\dfrac{x - 4}{x - 8}$ is not equal to $\dfrac{1}{2}$.

20-25. Simplify:

20. $\dfrac{5 \cdot 4 \cdot 3 \cdot 2 \cdot 1}{3 \cdot 2 \cdot 1}$

21. $\dfrac{4 \cdot 3 \cdot 2 \cdot 1}{5 \cdot 4 \cdot 3 \cdot 2 \cdot 1}$

22. $\dfrac{1 + 2 + 3 + 4 + 5}{1 + 2 + 3 + 4}$

23. $\dfrac{1 \cdot 2 \cdot 3 \cdot \ldots \cdot 10}{1 \cdot 2 \cdot 3 \cdot \ldots \cdot 12}$

24. $\dfrac{n\,(n - 1)}{n} \cdot \dfrac{6}{n - 1}$

25. $\dfrac{y}{x\,(x + 3)} \cdot \dfrac{(x - 3)\,x}{y}$

Lesson 7 (Optional)

Connecting Rate and Area

There are four basic models for multiplication and four for division. They are listed here. (The repeated subtraction model for division was mentioned only in the exercises on page 178.)

Multiplication	Division
repeated addition	repeated subtraction
ordered pair	splitting up
area	rate
size change	scale comparison

The "tree picture" of the ordered pair model can be used backwards as a splitting up model. The size change and scale comparison models are obviously related. So are the repeated addition and sub-traction models. Only the area and rate models seem unrelated.

249

But consider the following problem. You travel for 3 1/2 hours at a constant rate of 50 miles per hour. The distance travelled is easy to calculate:

$$50 \frac{\text{miles}}{\text{hr}} \cdot 3\frac{1}{2} \text{ hrs} = 175 \text{ miles}$$

But is there any way to picture this? The answer is "Yes" and the area model applies.

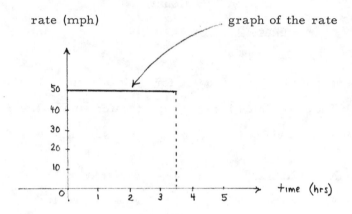

rate (mph)   graph of the rate

$$\text{Area} = \text{length} \cdot \text{width} = 50 \frac{\text{miles}}{\text{hr}} \cdot 3\frac{1}{2} \text{ hrs}$$

$$= 175 \text{ miles}$$

(Recall that in area one thinks of multiplying units. This works here.)

Now suppose you drive for another 1 1/2 hours at 55 mph.

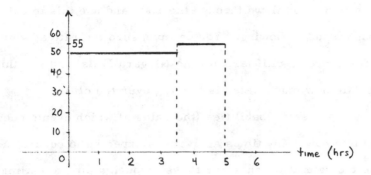

The distance can still be found by area. But now there are two areas
to be added. The right rectangle has area 82. 5 miles and so the
total distance driven is 257. 5 miles.

In actual real-life situations, it is impossible to keep the rate
constant. So the rate will not be described by segments but by a
curve. The area of the shaded portion under the curve gives the
distance.

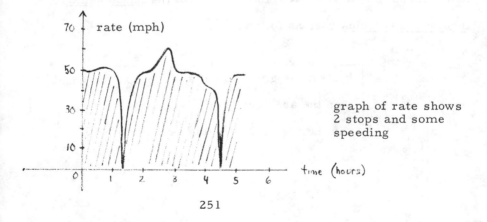

graph of rate shows
2 stops and some
speeding

251

You may wonder why anyone would want to use rate to calculate distance. It's usually the other way around. But there are many situations in which the rate is known and not distance or some other quantity. For example, you can measure the rate at which blood flows through the body and use this to calculate how much blood is flowing. You can measure the rate at which water flows over a cliff (as with the Niagara Falls) and use this to calculate how much water is flowing over the cliff.

You can use probabilities (the rates at which events occur) to calculate how many times an event is expected to occur. You can use the rate at which water flows through a pipe to estimate how much water is flowing through. These and many other quantities would be difficult or impossible to estimate without knowing the rates.

Because of these many situations, the calculation of area below curves is a fundamental idea in applications as well as in mathematics. This idea is normally first studied in the course called calculus taken by some high school seniors but more commonly found in colleges. A statement called the fundamental theorem of calculus goes as follows:

> If a rate is given in units y per x, then the area between a rate curve and the x-axis is the total amount of y's.

Example: The odometer of a car is the gauge that measures miles
          travelled. Suppose it breaks down and you are on a trip
          in a wilderness area. How can you determine how far
          you have travelled?

Solution: Keep track of the speed you are going. For instance,
          you might record it every 5 minutes. Then graph the
          points. Connect the points to form a "speed curve."
          The area under the speed curve will approximate the
          number of miles travelled. This area can be approx-
          imated by adding up the areas of the dotted rectangles.

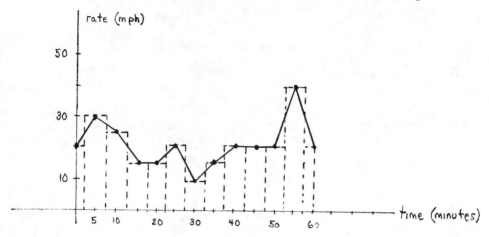

Questions covering the reading

1-4. Tell whether a multiplication or division model is named.
Then give the corresponding model in the other operation.

1.   repeated subtraction   2. size change

3.   area   4. cartesian product

253

5.   At right is the graph
     of the speed at which
     a car travelled.
     Calculate the total
     distance travelled.

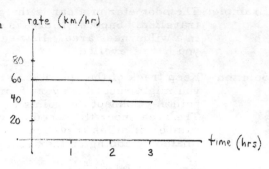

6.   Question 5 shows the relationship between what two models?

7.   Name a situation in which you might be able to calculate the
     rate y per x but in which you would not know the total amount
     of y's.

8.   In what course would you probably first study calculation of
     the area below a curve? (The area of a circle is an exception.)

9.   What is the fundamental theorem of calculus?

10. What is the odometer of a car?

11. How could you calculate how far you went on a driving trip if
     your speedometer was working but your odometer had broken?

Questions testing understanding of the reading

1.   Name a situation not mentioned in this lesson in which you
     might be able to calculate the rate y per x but in which
     you would not know the total amount of y's.

2. A car travels at the following rate for an hour. Estimate the distance traveled.

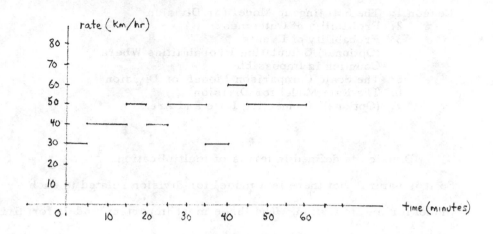

3. A place has the following probabilities of rain for a day in each month of the year. Consider each month to have 30 days. Use this information to estimate how many days of rain this place might be expected to have during a year.

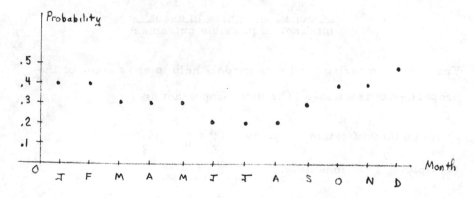

Division is defined in terms of multiplication: $\frac{a}{b} = a \cdot \frac{1}{b}$.
So it is natural that there is a model for division related to each
model for multiplication. The three most important models for division are in the lesson titles listed above.

One application of the splitting-up model is in the calculation
of __probability__. If all outcomes are assumed to be equally likely,
the probability of an event is:

$$\frac{\text{no. of outcomes in event}}{\text{total no. of possible outcomes}}$$

The scale comparison and rate models help to show some of the
properties of fractions. The most important are:

Fraction Simplification Property: $\frac{ka}{kb} = \frac{a}{b}$

Multiplying Fractions Rule: $\frac{a}{b} \cdot \frac{c}{d} = \frac{ac}{bd}$

Fractions and Opposites: $\frac{a}{-b} = \frac{^-a}{b} = -\frac{a}{b}$ and $\frac{^-a}{-b} = \frac{a}{b}$

CHAPTER 6

SENTENCE-SOLVING

Lesson 1

Estimating Wildlife Populations

People hunt rabbits and fish for food or sport. Whalers want oil as well as food. Scientists study migration habits of birds. Conservationists worry about endangered species. The garment industry wants furs. Park and forest rangers want to know what animals are around.

These people need to know how many animals there are of a given type. But animals move around. How can you count birds? What about whales or fish? Rabbits are very small. Each species of animal presents its own problems of counting.

Because direct counting is difficult or impossible, indirect methods are used to <u>estimate</u> animal populations. One indirect method is called the <u>catch-recatch</u> or <u>marking</u> method. Here is how it might work if you wanted to know how many trout were in a certain lake.

Suppose you caught 50 trout and 7 of them had a distinctive red marking. Then $\frac{7}{50}$ is the relative frequency of a trout having that marking. It is natural to think that $\frac{7}{50}$ is close to the probability that a trout is marked.   That is, you would think that

$$\frac{7}{50} \approx \frac{\text{total number of marked trout}}{\text{total number of trout in the lake}}$$

But we don't know how many trout are in the lake.   So what is often done is to handmark some trout (usually with paint) in advance and return them to the lake.   If 75 trout are marked, then

$$\frac{7}{50} \approx \frac{75}{\text{total number of trout in the lake}}$$

Let T stand for the total number of trout in the lake.   Then T is approximated by the solution to this equation.

$$\frac{7}{50} = \frac{75}{T}$$

Can you see that T is greater than 500 but less than 550?

In general, the marking method is as follows.   We wish to find **T, the** total number of trout.   (1) Mark m trout and return them to the lake.   (2) Catch C trout a few days later.   Let C_m be the total caught that were marked earlier.   (3) Assume

$$\binom{\text{relative frequency that}}{\text{caught trout is marked}} = \binom{\text{probability--assuming random-}}{\text{ness--that trout is marked}}$$

258

That is,

$$\frac{\text{no. of marked trout caught}}{\text{no. of caught trout}} = \frac{\text{total no. of marked trout}}{\text{total no. of trout}}$$

$$\frac{C_m}{C} = \frac{m}{T}$$

In this process, C_m, C, and m are known. T has to be found.

This is an example where finding the solution to an equation can help solve a real problem. But how do you solve this equation?

$$\frac{7}{50} = \frac{75}{T}$$

The next lessons show how this is done.

Questions covering the reading

1-8. Why might each of the following people be interested in wild-life populations?

1.   whaler   2.   fisherman

3.   tourist   4.   scientist

5.   furrier   6.   conservationist

7.   Suppose you want to count the number of deer in a particular area of a forest. Describe how the marking method might be used.

259

8. In the marking method, what number is assumed equal to the relative frequency that a caught animal is marked?

9. Suppose 200 trout are marked and replaced in a lake. Later 150 trout are captured and 3 of these are found to be marked. If T is an estimate of the total number of trout, how might T, 200, 150, and 3 be related?

Questions testing understanding of the reading

1-2. In 1970 in Dryden Lake, New York, a marking estimate was done. The first fish were marked on their fins in mid-November. The second catch of fish was done around December 1st. (Fish do not enter this lake from outside sources.) Here is the data that was collected. (For more problems like this, see Question for discussion #6, p. 262.)

	Marked (Nov.)	Captured (Dec.)	Captured and Previously Marked (Dec.)
Large-mouth bass	213	104	13
Pickerel	232	329	16

Give an equation which could be solved to estimate the number of:

1. large-mouth bass in the lake.   2. pickerel in the lake.

3-5. Suppose you wish to estimate t, the total number of marbles in a large bag. So you take 10 marbles out and paint a spot on them, let the paint dry, and return the marbles. You mix the marbles and then later take out 25 marbles. Of those m are marked.

3. What equation estimates a relationship between m, t, 10, and 25?

4. Estimate t if m = 3.   5. Estimate t if m is 2.

6. St. Paul Island in Alaska has 12 fur seal rookeries (breeding places). In 1961, to estimate the fur seal pup population in Gorbatch rookery, 4963 fur seal pups were tagged in early August. In late August a sample of 900 pups was examined and 218 of these were previously tagged. (This data is from the July, 1968, issue of the <u>Transactions of the American Fisheries Society</u>.) Let F be the number of fur seal pups in Gorbatch rookery. Solving what equation gives an estimate for F?

7. Suppose that you have a large number of paper clips. Describe how the marking method could be used to estimate the total number of paper clips.

8. <u>Multiple choice.</u> In a forest there are D deer and you wish to estimate D. You capture and mark 100 deer. Later you catch 50 deer and all 50 are marked. Which of choices (a)-(d) is <u>not</u> possible?

    (a) D = 100   (b) D = 50   (c) D = 300
    (d) D = 5000   (e) All are possible

9. The marking method does not lead to the exact population. It only gives an estimate. Why is this not a particular weakness of the method?

Questions for discussion or exploration

1-4. The marking method makes some assumptions not mentioned in the lesson. How would you try to assure that each of the following happens? Can you make sure that these things happen?

1. The marked animals are not affected by marking and the marks or tags do not come off.

2. The marked animals are mixed in the population.

3. The non-marked animals are just as available for capture as the marked animals.

4. The population of the animals does not change between marking and recapture.

5.   Create an experiment at home which uses the marking method.
       (For example, you might try to estimate the number of hair-
       pins or paper clips or rubber bands or nails in a place where
       large numbers of these are kept.)

6.   For more information about estimating wildlife populations,
       read "Estimating the Size of Wildlife Populations," by
       S. Chatterjee, in Statistics by Example (Reading, MA: Addi-
       son-Wesley, 1973), pp. 99-104.

7.   For information on how whale populations are estimated,
       read "The Plight of the Whales," by D.G. Chapman, in
       Statistics: A Guide to the Unknown (San Francisco: Holden-
       Day, 1972), pp. 84-91.

Lesson 2

The Multiplication Property of Equations

We begin with a short review. A sentence with a variable
in it is an open sentence.

$$\frac{7}{50} = \frac{75}{T}$$ is an open sentence

A value of the variable which makes the sentence true is called a
solution. Finding this value is called solving the sentence.

Each equation below has solution 5. (Check them)

$$2x = 10 \qquad \frac{x + 47}{13} = x - 1 \qquad -3 = x - 8$$

We call these equations equivalent because they have the same
solution.

Definition:

> Two open sentences are <u>equivalent</u> if and only if every solution to one is a solution to the other.

It is easy to create equivalent sentences. Here is an equation whose solution is 5.

$$2x = 10$$

Now multiply both sides by -4. (The abbreviation M_{-4} tells what was done.)

$$M_{-4}: \quad -4 \cdot (2x) = -4 \cdot 10$$
$$-8x = -40$$

The solution is still 5. The sentences are equivalent. This illustrates a very important property of multiplication. It will be helpful in sentence-solving.

<u>Multiplication Property of Equations</u>:

> If $x = y$, then $ax = ay$.

There is a way of visualizing this property. Think of a figure
(like the one below at left) which contains two segments of lengths
x and y. Now apply the scale factor _a_ to the figure.

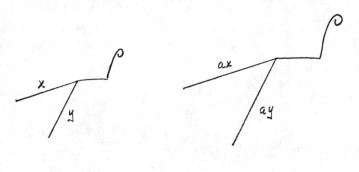

original   image, _a_ times the size

The multiplication property of equations now states the obvious:
If the lengths are equal at left then the corresponding lengths are
equal at right. That is, if x = y, then ax = ay.

There is also an arithmetic way of thinking about the multiplication property of equations. Here are two expressions which
look different but are equal.

$$.75 = \frac{3}{4}$$

Suppose both sides are multiplied by 6. Will the answers be equal?
(Notice the question mark above the equal sign.)

$$M_6: \qquad\qquad 6 \cdot .75 \overset{?}{=} 6 \cdot \tfrac{3}{4}$$

$$4.50 \overset{?}{=} \tfrac{18}{4}$$

Change to fractions: $\quad 4\tfrac{1}{2} \overset{?}{=} \tfrac{9}{2}$ $\quad$ Of course they are equal.
We just wrote them in a
different language.

Questions covering the reading

1-3. Define:

1. open sentence. $\qquad\qquad$ 2. solution to an open sentence.

3. equivalent sentences.

4. What have you done when you have "solved" a sentence?

5-6. Suppose $x = 8$. **True** or **False**?

5. $3x = 38$ $\qquad\qquad\qquad$ 6. $3x = 24$

7-8. Suppose $2y = \pi$. True or False?

7. $8(2y) = 8\pi$ $\qquad\qquad$ 8. $-2y = -\pi$

9. What is the Multiplication Property of Equations?

10-13. It is true that $\tfrac{1}{4} = .25$ and $\tfrac{18}{3} = 6$. True or False?

10. $-7 \cdot \tfrac{1}{4} = -7 \cdot .25$ $\qquad$ 11. $\tfrac{1}{4} \cdot \tfrac{18}{3} = .25 \cdot 6$

12. $\tfrac{18}{3} \cdot a = 6a$ $\qquad\qquad$ 13. $\tfrac{11}{2} \cdot \tfrac{18}{3} = \tfrac{11}{2} \cdot 6$

14. A drawing has two segments of the same length. Suppose you double the size of this drawing. Will the corresponding segments in the larger drawing have the same length?

15-18. The abbreviation M_n means "multiply each side by n."
Begin with the sentence $-3x = 42$. It has the solution -14.
(a) What sentence results from doing the indicated action?
(b) Is -14 a solution to your answer to (a)?

15. M_5 $\qquad$ 16. M_3 $\qquad$ 17. M_{-10} $\qquad$ 18. $M_{-\frac{1}{3}}$

19-22. The sentence $40 = \frac{D}{30}$ has the solution 1200. (a) What sentence results from doing the indicated action? (b) Is 1200 a solution to your answer to part (a)?

19. M_{-40} 20. M_{30} 21. $M_{\frac{1}{40}}$ 22. M_{60}

Questions applying the reading

1-4. Apply the given action to the given sentnece.

1. $5x = 110$; $M_{\frac{1}{5}}$ 2. $ab = c$; $M_{\frac{1}{a}}$

3. $7 = \frac{2y}{3}$; $M_{\frac{3}{2}}$ 4. $6t = 12$; M_{-6}

5-10. An idea similar to the multiplication property of equations can be applied in some conversion questions.

Example: How many cm equal 1 foot?

Solution: We know   2.54 cm = 1 in.
            M_{12}:   12 · 2.54 cm = 12 · 1 in.
                             30.48 cm = 12 in. = 1 ft.

(This idea works with all the conversion formulas you have had except those for temperature. It doesn't work when 0 in one unit is not 0 in the other unit.)
Apply this idea to estimate:

5.   the number of cm in a yard.

6.   the number of cm in a mile.

7.   the number of pounds in 8 kg.

8.   the number of pounds in .3 kg.

9.   the number of km in 1/8 of a mile. (Recall 1.6 km ≈ 1 mi.)

10.   the number of km in 10 miles.

11-14. The multiplication property of equations has been applied to get sentence (ii) from sentence (i). By what has each side of sentence (i) been multiplied?

11. (i) $9A = 20$ (ii) $3A = \frac{20}{3}$

12. (i) $\frac{B}{6} = 11$ (ii) $B = 66$

13. (i) $-1 = \frac{1}{2}C$ (ii) $-2 = C$

14. (i) $.1D = 867$ (ii) $D = 8670$

15-18. Do these problems without finding a solution to the given equations.

15. Suppose $30x = \frac{2}{5}$. Then $60x =$ _____.

16. Given $4y = .039$. Then $4000y =$ _____.

17. Given $6z = 10$. Then $3z =$ _____.

18. Given $8A = 5B$. Then $40A =$ _____.

Lesson 3

An Algorithm for Solving $ax = b$

An **algorithm** is a procedure for solving a certain type of problem. Algorithms are found inside and outside of mathematics. For example, if the problem is to bake a cake, a recipe is an algorithm for doing this. Directions for building a radio form an algorithm. You are able to multiply fractions easily because you know an algorithm for this.

It is useful to have algorithms for finding solutions to sentences. You only need to learn a few algorithms. We begin with

267

equations of the form ax = b. If we want to solve this equation for x, we call x the <u>unknown</u>. We call a the <u>coefficient</u> of x. (If two numbers are multiplied, each number is the coefficient of the other.)

Here is a typical example.

$$\frac{2}{3}m = 40 \qquad m \text{ is the unknown.}$$
$$\frac{2}{3} \text{ is the coefficient of m.}$$

You can multiply both sides of the equation by any number. In this case, it is wise to multiply by $\frac{3}{2}$.

$$M_{\frac{3}{2}}: \qquad \frac{3}{2} \cdot (\frac{2}{3}m) = \frac{3}{2} \cdot 40$$

$$(\frac{3}{2} \cdot \frac{2}{3})m = 60 \qquad \text{(by the associative property)}$$

$$1 \cdot m = 60 \qquad \text{(reciprocals multiply to 1)}$$

$$m = 60 \qquad \text{(1 is the identity for multiplication)}$$

The number $\frac{3}{2}$ was picked because it is the reciprocal of $\frac{2}{3}$. Multiplying those numbers together causes 1 to become the coefficient of m. By doing this, the last equation has m alone on one side. And 60 is the obvious solution to the last sentence. So it works in the given sentence. $\frac{2}{3} \cdot 60 = 40$.

All equations of this form are solved this way.

$$\boxed{\text{Algorithm 1: To solve } ax = b \text{ for } x, \text{ multiply both sides of the equation by } \tfrac{1}{a}, \text{ the reciprocal of a.}}$$

Examples: Solving equations of the form $ax = b$.

1. $-4x = \dfrac{2}{3}$ Think: x is unknown, -4 is the coefficient of x.

Multiply both sides by $-\dfrac{1}{4}$. Then simplify.

$M_{-\frac{1}{4}}:$ $\qquad -\dfrac{1}{4} \cdot (-4x) = -\dfrac{1}{4} \cdot \dfrac{2}{3}$

$\qquad\qquad -\dfrac{1}{4} \cdot -4 \cdot x = -\dfrac{2}{12}$

$\qquad\qquad\qquad x = -\dfrac{1}{6}$

2. $\dfrac{2}{3} = \dfrac{v}{19}$ Think: v is unknown, $\dfrac{1}{19}$ is the coefficient of v.

$M_{19}:$ $\qquad\qquad 19 \cdot \dfrac{2}{5} = 19 \cdot \dfrac{v}{19}$

$\qquad\qquad\qquad \dfrac{38}{5} = v$

(Notice how easy it is. You can check the answer by substitution in the original equation.)

3. In driving, mean rate $= \dfrac{\text{distance travelled}}{\text{total time}}$

$$r = \dfrac{d}{t}$$

To solve for d: Think: $\dfrac{1}{t}$ is the coefficient of d. So multiply by t.

$M_t:$ $\qquad\qquad tr = t \cdot \dfrac{d}{t}$

$\qquad\qquad\qquad tr = d$

This formula is usually written $\underline{d = rt}$. You should learn it.

Look again at the algorithm. In the last chapter you learned that dividing by $\underline{a}$ is the same as multiplying by $\frac{1}{a}$. This means that the algorithm can be reworded.

Algorithm 1 (reworded): To solve ax = b for x, divide both

sides of the equation by a.

The reworded version is useful when decimals are involved.

4.   Solve for P: $1.03P = 150$

Think: 1.03 is the coefficient of P. Multiplying by $\frac{1}{1.03}$ is the same as dividing by 1.03.

$$\frac{1.03P}{1.03} = \frac{150}{1.03}$$

$$P = 145.63\ldots$$

Again you can check by substitution in the original sentence.

$$1.03(145.63) = 149.9989, \text{ which is close enough.}$$

Questions covering the reading

1-8. For each sentence: (a) Name the unknown. (b) Name the coefficient of the unknown. (c) Solve the sentence.

1. $3v = 8$ 2. $\frac{2}{3} = \frac{x}{4}$

3. $17 = -18y$ 4. $1.3x = 39$

5. $\frac{2t}{5} = \frac{4}{35}$ 6. $100000W = 1$

7.   Solving for d: $r = \frac{d}{t}$ 8.   Solving for x: ax = b

9. What is the algorithm for solving $ax = b$?

10. What is an alternate algorithm for solving $ax = b$?

11. Define: algorithm.

12. Give an example of an algorithm found outside of mathematics.

13-22. (a) Solve. (b) Check your answers by substituting in the original equation.

13. $80x = 7280$ 14. $13y = 18.2$

15. $-\dfrac{w}{3} = 10$ 16. $\dfrac{1}{3} = -2a$

17. $4.3 = \dfrac{x}{100}$ 18. $r \cdot 2\dfrac{1}{2} = 6$

19. $x \cdot 9 = -12$ C 20. $23.1t = .77$

21. $\dfrac{m}{6} = \dfrac{2}{5}$ 22. $\dfrac{5}{7} = \dfrac{n}{112}$

23. With consistent units, there is a formula $r = \dfrac{d}{t}$. What do r, d, and t stand for? What is the more usual way of writing this formula?

24. Solve for **m**: $F = ma$ 25. Solve for a: $F = ma$

26. Solve for h: $A = \dfrac{1}{2}bh$ 27. Solve for b: $A = \dfrac{1}{2}bh$

28. How could you check your answers to Questions 24-27?

29-34. More practice. Solve and check.

29. $.75A = 4$ 30. 75% of B is 1464

31. $\dfrac{x}{6} = -3$ 32. $9y = 3$

33. $-19k = -18$ 34. $4 = \dfrac{3}{5} \cdot m$

Questions applying the reading

1-6. In banking, the formula $\underline{I = prt}$ relates simple interest I, principle (how much is invested) p, rate r, and time t. This formula is correct if you do not invest the interest you earn.

1.   If you invest \$500 for 2 years, what rate do you need in order to earn \$30 in interest?

C   2.   If you want to earn \$100 interest in 1 year, how much do you need to invest if the rate is 6%?

C   3.   Repeat Question 2 if the rate is 7%.

4.   How long would it take to earn \$2000 if you invested \$5000 at 8%?

5.   Solve the formula for r.

6.   Solve the formula for p.

7-8. The safe working strength S of a leather belt (in pounds) is approximated by the formula $S = 300\,wt$, where w is the width and t its thickness (in inches).

7.   If a belt is $\frac{1}{16}$" thick, how wide must it be to lift 25 lbs?

8.   If a belt is 3" wide, how thick must it be to lift 40 lbs?

9-10. The formula $A = \frac{1}{2}bh$ relates the area, height, and length of the base of a triangular region.

9.   How high must a triangle be to have an area of 20 sq cm if its base has length 5 cm?

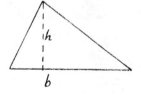

10.   Repeat Question 9 if the base is a meter long. What would such a triangle look like?

11-16.   Find the solution <u>to the nearest tenth.</u> Try to do these without a calculator.

11. $3.0001x = 6$
                        12. $\frac{v}{7.002} = 14$

13. $\frac{1}{3} = 11.99998W$ 14. $22t = 439.99999992103$

15. $14.9999M = 15.0000009$ 16. $1.00008k = 12.34$

17. Is it possible to have a rectangle with length 10m and area 1 sq m? If so, what is the width?

18. At an outdoor band concert, 60 people can be seated in each row. Suppose there are r rows. What is r if a seating capacity of 2500 is desired?

19. In physics, the formula $I = \frac{E}{R}$ is used. (a) Look in a physics book to find out what E, I, and R stand for. (b) Solve the equation for E.

C 20. How long should it take to drive from Los Angeles to San Francisco, a distance of 403 miles, if you travel at 55 mph and allow 1 hour for stops? (Use the formula $d = rt$.)

Questions for discussion

1. The word "algorithm" is named after the foremost Islamic scholar of the 9th century, Mohammed ibn-Musa al-Khowarizmi. (His first names mean "Mohammed, the son of Musa.") Look in an encyclopedia to find out some of the accomplishments of this man.

2. Why do you think the length has nothing to do with the strength of a belt?

3. The algorithm for solving $ax = b$ does not work when $a = 0$. Why not?

4. To show in detail that $x = \frac{11}{9}$ follows from $9x = 11$, a student went through the following process. What property or definition was used in each step?

$$9x = 11$$

(a) $$\frac{1}{9} \cdot (9x) = \frac{1}{9} \cdot 11$$

(b) $$(\frac{1}{9} \cdot 9)x = \frac{1}{9} \cdot 11$$

(c) $$1 \cdot x = \frac{1}{9} \cdot 11$$

(d) $$x = \frac{1}{9} \cdot 11$$

(e) $$x = \frac{11}{9}$$

273

5. Refer to Questions applying the reading #2, 3, and 6. Would doing #6 have saved some steps in doing #2 or #3?

Lesson 4

Proportions

In Lesson 1, counting trout in a lake **led** to the equation $\frac{7}{50} = \frac{75}{T}$. The multiplication property of equations can help solve this equation. The idea is to multiply both sides by a number picked to get rid of fractions. In this case, multiply by 50T. Look at what happens.

Given: $$\frac{7}{50} = \frac{75}{T}$$

$$M_{50T}: \quad 50T \cdot \frac{7}{50} = \frac{75}{T} \cdot 50T$$

Now simplify. The resulting equation has no fractions.

$$7T = 3750$$

This is an equation you know how to solve.

$$M_{\frac{1}{7}}: \quad T = \frac{3750}{7}$$
$$= 535\frac{5}{7}$$

An estimate for the number of trout in the lake is 536.

An equation of the form $\frac{a}{b} = \frac{c}{d}$ is called a <u>proportion</u>. Many situations lead to proportions.

Example 1:   A motorist decides to keep records of gasoline mileage. The car was able to go 216 miles on 13.8 gallons of gas. At this rate, how far can the car go on a full tank of 21 gallons?

Solution: Let f be the distance it can go on a full tank. The rate of using gas is then

$$\frac{f \text{ miles}}{21 \text{ gallons}} \quad \text{or} \quad \frac{f}{21} \text{ miles per gallon.}$$

But the car's known rate of gas guzzling is

$$\frac{216 \text{ miles}}{13.8 \text{ gallons}} \quad \text{or} \quad \frac{216}{13.8} \text{ miles per gallon.}$$

If the rates are equal, then $\frac{f}{21} = \frac{216}{13.8}$. Now solve.

$$M_{21}: \qquad\qquad f = 21 \cdot \frac{216}{13.8}$$

Simplify: $f \approx 329$

The car can go about 329 miles on a full tank.

These examples show that you already know enough about solving equations to solve any proportion. But proportions are so common that a special property is used to shorten the work. The idea is as follows.

Suppose $\frac{a}{b} = \frac{c}{d}$.

Now multiply both sides by bd. The resulting equation has no fractions.

$$M_{bd}: \qquad bd \cdot \frac{a}{b} = bd \cdot \frac{c}{d}$$

Simplify: $ad = bc$

These steps have shown:

Means-Extremes
Property:

$$\text{If } \frac{a}{b} = \frac{c}{d}, \text{ then } ad = bc.$$

The means-extremes property gets its name from olden days.
Proportions used to be written using ratio notation.

$$a : b = c : d$$

The numbers a and d were called the <u>extremes</u> (outer terms);
b and c are the <u>means</u> (terms in the middle). Since ad = bc, the
<u>the product of the means equals the product of the extremes</u>.

Example 2:   If cans of orange juice are 3 for 86¢, how much
               will 5 cans cost?

Solution: Let c be the cost of 5 cans.   There are two
               equal rates.

$$\frac{3 \text{ cans}}{\$.86} = \frac{5 \text{ cans}}{c \text{ dollars}}$$

               Applying the Means-Extremes Property,

$$3c = 5 \cdot .86$$

$M_{\frac{1}{3}}$:

$$c = \frac{5 \cdot .86}{3} = \frac{4.30}{3} = 1.4\overline{3}$$

               The cost will be $1.44 unless there is a special
               rate for quantity.

276

Questions covering the reading

1. What is a proportion?

2-4. In each proportion, suppose you want to get rid of fractions. (a) Multiplying both sides by what number will do this? (b) What equation results?

2. $\dfrac{7}{50} = \dfrac{75}{T}$

3. $\dfrac{8}{-5} = \dfrac{x}{11}$

4. $\dfrac{a}{b} = \dfrac{c}{d}$

5. What is the Means-Extremes Property and how did it get its name?

6-8. A proportion is given. (a) Name the means. (b) Name the extremes. (c) Use the Means-Extremes Property to solve the proportion.

6. $\dfrac{27}{y} = \dfrac{54}{5}$

7. $\dfrac{z}{6} = \dfrac{-2}{5}$

8. $\dfrac{85}{14} = \dfrac{102}{x}$

9. Solve the equation of Question 7 without using the Means-Extremes property. Use only the multiplication property of equations.

10-11. A question is given. (a) Solving what proportion will answer the question? (b) Solve the proportion.

10. If 6 cans of pineapple juice cost \$1.69, what would 10 cans cost?

11. If a car goes 500 km on 70 liters of gas, how far will it go on a full tank of 100 liters?

12-14. You may be able to answer these questions in your head. Even if this is so, write down a proportion that will help answer the question.

12. If your heart beats 19 times in 15 seconds, how many times will it beat in 60 seconds?

13. If .7cm of rain falls in 3 hours, at the same rate how much will fall in 9 hours?

14. If you read 40 pages of a novel in 50 minutes, how long will it take you to read the entire 200-page novel (assuming the same rate)?

15. Why is the assumption in Question 14 probably not a good one to make?

277

Questions testing understanding of the reading

1. Order these terms from most general to most specific.

   proportion   equation   sentence

C 2. On the first three days of the hunting season in Deer County, 79 deer were bagged. (a) If the hunting season is 10 days long, and the bagging continues at this rate, how many deer will have been bagged in the season? (b) How good do you think this estimate is?

C 3. The National Safety Council identifies some holiday weekends as beginning at 6 PM Friday and lasting until midnight Sunday. If 147 people in the U.S. are killed by 6 PM Saturday, how many people will die on highways in the entire weekend (assuming the rate continued)?

4-5. These proportions come from the actual wildlife situations in Lesson 1. Solve the proportion and tell what your answer means.

C 4. $\dfrac{218}{900} = \dfrac{4965}{t}$ (Question 6, p. 261)

C 5. $\dfrac{16}{329} = \dfrac{232}{t}$ (Question 2, p.260)

6-8. 8 km ≈ 5 miles. Fill in the blanks.

6. 7 km ≈ _____.   7. _____ km ≈ 12 miles.

8. 10 km ≈ _____ miles.

9. A recipe for 4 people calls for $1\frac{1}{2}$ tsp. salt. For 7 people, how many tsp. should be used?

10. A recipe for 3 people calls for $\frac{1}{2}$ liter of water. For 5 people, how many liters should be used?

11-12. 110 yards ≈ 100 meters. Then:

11. 400 yards ≈ ___ meters.   12. 3 meters ≈ ____ yards.

13. Suppose a pitcher in baseball allows 10 runs in 15 innings. At this rate, how many runs would be allowed in 9 innings? (The answer to this question is the pitchers "earned run average," or ERA.)

14-15. Refer to Question 13 for an explanation of ERA.

14. What is the ERA of a pitcher who allows 17 runs in 49 innings?

15. What is the ERA of a pitcher who allows 17 runs in 58 innings?

16. A basketball team scores 17 points in the first 6 minutes of play. At this rate, how many points would it score in a 32-minute game?

17-20. The Means-Extremes Property can tell whether fractions are equal. $\frac{a}{b}$ and $\frac{c}{d}$ will be equal only if ad = bc. Tell whether the given fractions are equal.

17. $\frac{1}{3}$, $\frac{33}{100}$

18. $\frac{1001}{65}$, $\frac{693}{45}$

19. $\frac{41x}{29}$, $\frac{29x}{17}$

20. $\frac{4.5}{-5}$, $\frac{-153}{170}$

21. A family decided to keep track of the number of phone calls made. In the first week of December, 38 calls were made. At this rate, how many will be made for the month?

22. Make up a question of your own which could be answered by solving a proportion.

Question for discussion

1. Suppose the cost of a 10" pizza is $3.50. Then the cost of a 14" pizza should be $4.90, because

$$\frac{\$3.50}{10"} = \frac{\$4.90}{14"}.$$

What is wrong with this reasoning?

Lesson 5

An Algorithm for $a + x = b$

Suppose your age is x and a friend's age is y. Five years from now your age will be $x + 5$, your friend's age is $y + 5$. If the two of you are the same age now, then

$$x = y$$

Then five years from now the ages will still be the same.

$$x + 5 = y + 5$$

And $\underline{a}$ years from now the ages will be the same.

$$x + a = y + a$$

This example illustrates the Addition Property of Equations.

Addition
Property of
Equations:

> If $x = y$, then $a + x = a + y$.

The **Addition** Property of Equations can be applied to help solve equations. Consider the following equation.

$$^-8 + x = 43$$

You can add any number to both sides. In this case it is wise to add 8.

280

The abbreviation A_8 is used.

$$A_8: \qquad 8 + (-8 + x) = 8 + 43$$
$$(8 + -8) + x = 51 \qquad \text{(Associative property)}$$
$$0 + x = 51 \qquad \text{(Property of opposites)}$$
$$x = 51 \qquad \text{(0 is the identity for addition)}$$

The number 8 was picked because it is the opposite of -8. Adding these numbers gives 0, so x is alone on the left side. This idea works in all such cases.

Algorithm 2: To solve $a + x = b$ for x, add $-a$ to both sides of the equation.

Examples: Solving equations of the form $a + x = b$.

1. $\underline{1.2 + m = 6}$ \qquad Think: m is unknown, 1.2 is added.

Add -1.2 to both sides. Then simplify.

$$A_{-1.2}: \qquad -1.2 + (1.2 + m) = -1.2 + 6$$
$$-1.2 + 1.2 + m = 4.8$$
$$0 + m = 4.8$$
$$m = 4.8$$

2. $\underline{x - 703 = -112}$ First convert the subtraction to addition.

$$x + -703 = -112$$

Now think: x is the unknown. -703 is added to it. So add

703 more to get x alone.

A_{703}: $703 + x + -703 = 703 + -112$

$$x = 591$$

3.   The perimeter p of a triangle

with sides a, b, and c satis-

fied the formula

   $\underline{p = a + b + c.}$

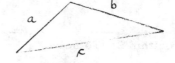

If two sides of the triangle have

lengths 42 and 106 and the perimeter is 250, calculate the

length of the third side.

Solution: (You could do this in your head but here is another

way.)

From the given, p = 250, a = 42, b = 106.

Substituting,   250 = 42 + 106 + c

250 = 148 + c

A_{-148}:   102 = c

The third side has length 102.

Questions covering the reading

1. What is the Addition Property of Equations?

2-5. Suppose your age is x and a friend's age is y. What does each sentence mean?

2. $x = y$ 3. $x + 5 = y + 5$

4. $x - 13 = y - 13$ 5. If $x = y$, then $x + a = y + a$.

6-9. Apply the given operation to both sides of the equation $42 + x = 59$. Then simplify.

6. A_{42} 7. A_{59} 8. $M_{\frac{1}{42}}$ 9. A_{-42}

10. Which of Operations 6-9 is most helpful in solving the equation?

11-14. Apply the given operation to both sides of $a + x = b$. Then simplify.

11. A_{-a} 12. $M_{\frac{1}{a}}$ 13. A_{-b} 14. A_{-x}

15. Which of Operations 11-14 is most helpful in solving the equation for x?

16. Which of Operations 11-14 is most helpful in solving the equation for a?

17-22. Solve each equation and check your answer.

17. $89 = 97 + x$ 18. $A + 3 = -8$

19. $-\frac{13}{5} = y + \frac{1}{5}$ 20. $B + -6 = 41.2$

21. $1 + k = -1$ 22. $-10 = m + -30$

23-26. Let p be the perimeter of a triangle. Let x, y, and z be the lengths of the sides of the triangle. (Refer to Example 3, p. 282.)

23. Calculate x if $p = 100$, $y = 30$, and $z = 26$.

24. If $x = 1.29$ cm, $y = 1.68$ cm, and $p = 3.43$ cm, what is z?

25. Find the length of the third side if two sides have lengths $\frac{1}{2}$ and $\frac{1}{3}$ and the perimeter is 1.

26. Find y if p = 746, x = 294, and z = 274.

27-32. Solve and check. (You may find it helpful first to convert the subtractions to additions.)

27. a - 1.3 = .6

28. b - -24 = 12

29. -25 = x - 4

30. $\frac{1}{2} = y - \frac{1}{2}$

31. m - 73 = -862

32. n - 1.42 = -1.2

Questions testing understanding of the reading

1-6. These problems require more complicated arithmetic.

1. $2 + x = \frac{3}{7}$

2. n - 163 = 359

3. 14 = 6 + n + 80

4. -11 + -11 = -12 + d

5. 0.083 + t = 0.947

6. $u + \frac{3}{5} = \frac{1}{3}$

7-10. The probabilities that an event <u>does</u> occur and does <u>not</u> occur are related by the formula d + n = 1.

7. Calculate n if $d = \frac{2}{3}$.

8. If the precipitation probability is 35%, what is the probability that there will be no rain or snow?

9. Suppose $\frac{33}{70}$ of people polled feel that the government is spending too much money on defense. What fraction of people do not feel this way?

10. Solve the formula for d.

11-14. The formula P = C + I + G describes U.S. production P, consumption C, investment I, and portion G used by government (all variables in dollars). It is an economic formula. (Source: "America the Mathematized," by Harry Schwartz, <u>New York Times</u>, Nov. 7, 1969.)

11. Calculate I if P is 800 billion dollars, C is 500 billion dollars and G is 200 billion dollars.

12. Solve the formula for C.   13. Solve the formula for I.

14. Solve the formula for G.

15. Two children wish to buy a present for their parents which costs $50. If one child has saved s dollars, how much does the other child need to have saved?

C 16. A husband earned h dollars last year. His wife earned w dollars. Suppose their income tax form shows that they earned $21,349.27 in all. (a) What equation relates these three numbers? (b) If the wife claims to have earned $10,684.39, how much was earned by the husband?

17-18. In geometry, two angles with measures x and y are called <u>complementary</u> if x + y = 90.

17. Solve this equation for x.   18. If y = 29, what is x?

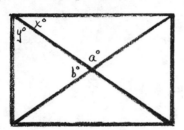

On this drawing of a gate, x + y = 90 and a + b = 180.

19-20. In geometry, two angles with measures a and b are called <u>supplementary</u> if a + b = 180.

19. If one angle has measure 41.2, what is the measure of a supplementary angle?

20. Solve for b.

21. If the temperature is -11° C, by how much must it increase to become 13° C? (a) Do this problem in your head. (b) Imagine that you couldn't do the problem in your head. Write an equation which could help answer the question.

285

1. To show in detail that $x = {}^-8$ follows from $x + 10 = 2$,
    a student went through the following process. What prop-
    erty of definition was used in each step?

$$x + 10 = 2$$

(a) $(x + 10) + {}^-10 = 2 + {}^-10$ (?   )

(b) $x + (10 + {}^-10) = 2 + {}^-10$ (?   )

(c) $x + \quad 0 \quad = 2 + {}^-10$ (?   )

(d) $x = 2 + {}^-10$ (?   )

                  $x = {}^-8$ (arithmetic)

Lesson 6

Inequalities

1. You have $20 to buy tickets for a
    concert. If tickets are $4 each,
    how many can you buy?

Most people would answer 5 to this question. But a more accurate

answer is 0, 1, 2, 3, 4, or 5. 5 is the largest number of tickets

you can buy.

Let us examine this problem mathematically. Because each

ticket is $4, n tickets would cost 4n dollars. Since you may spend

any amount less than or equal to 20 dollars,

$$4n \leq 20$$

286

The nonnegative integer solutions to this sentence are the answers to the original question.

The sentence $4n \leq 20$ is an example of an _inequality_. An inequality is any sentence which contains the symbols $<$, $\leq$, $>$, or $\geq$. Many real situations are more accurately described by inequalities than by equations.

2. A person was caught speeding in a 30 mph zone. If the police allow a 5 mph leeway, describe the car's possible speed S.

Answer: $S > 35$

The sentence $S > 35$ has infinitely many solutions. So solutions cannot be listed. It is easier to _graph_ them. Here is the graph of the solution set to $S > 35$.

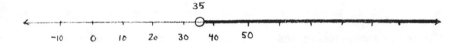

The small circle at the number 35 indicates that 35 is not a solution. Inequalities are helpful in describing estimates.

3. It is estimated that 2,000,000 copies of a certain record were sold. This estimate is thought to be off by as much as 30%. Let r be the number of records sold. What sentence is satisfied by r?

Answer: Since 30% of 2,000,000 is 600,000:

$1,400,000 \leq r \leq 2,600,000$

287

Here is the graph of all solutions. (The scale is in millions.)

You have now learned three ways of showing solutions to sentences.

Ways of showing solutions	x + 10 = 45	4n ⩽ 20
1. List the solutions.	35	cannot be listed (infinitely many)
2. Write a very simple sentence which is equivalent to the given sentence.	x = 35	n ⩽ 5
3. Graph the solutions.		

In this book, we use all of the above ways. Occasionally we use a fourth way, and your teacher may prefer it over the others. It is to write the <u>solution set</u>. The solution set is just what the name tells you--it is the set of all solutions. Notice how solution sets are written.

Way of showing solutions	x + 10 = 45	4n ⩽ 20
4. Give the solution set.	$\{35\}$	$\{n: \ n \leqslant 5\}$

The symbol $\{n: \quad\}$ is read "the set of numbers n which satisfy..." Notice that in order to show the solutions to an inequality you need either to graph or to solve the inequality. Listing solutions is often impossible. So the next lessons cover how to solve inequalities.

Questions covering the reading

1. What symbols are found in inequalities?

2. You have $25 to buy tickets for a concert. Suppose tickets are $2.50 each.
   (a) What is the largest number of tickets that you can buy?

   (b) How many tickets can you buy?

   (c) Let n be the number of tickets you can buy. What sentence is satisfied by n?

3-8. (Review) Solutions to what sentences are pictured?

3.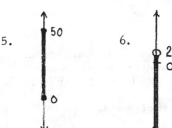

4.

5.

6.

7.

8.

9-14. (Review) Graph all solutions to each sentence.

9. x > 8.9

10. y < -1

11. $\frac{1}{3} \leq z \leq \frac{2}{3}$

12. -4 ≥ A

13. $\frac{2}{3} > B$

14. -40 < C ≤ -38

15. In words, what is $\{n: \quad\}$?

289

16. In words, what is $\{x: x > 5\}$?

17. True or False: The solution set to $\underline{2n = 14}$ is $\{7\}$.

18. True or False: The solution set to $\underline{2n \leq 14}$ is $\{n: n \leq 7\}$.

19-24. Write the solution set to each sentence in Questions 9-14.

25. What are the four ways of showing solutions to a sentence?

26. Show the solution to $\underline{5 + x = {}^-2}$ in four different ways.

27-29. Some interstate highways have a minimum speed of 45 mph (under normal conditions). Maximum speed is 55 mph. (a) Describe each situation by means of an inequality. (b) Graph all solutions.

27. A car is travelling too slow.

28. A car is speeding.

29. A car is going within the speed limits.

Questions applying the reading

1-8. Describe the situation by means of an inequality.

1. Automobiles are about 3 to 7 meters long. (Let L be the length of an automobile.)

2. Let E be the elevation (in meters) of a point in California. California elevations range from 86 meters below sea level (Death Valley) to 4429 meters above sea level (Mt. Whitney).

3. Her weight is within 1 kg of 53 kg. (Let w stand for her weight.)

4. Rounded to the nearest mm, that fish is 17 mm long. (Let L be the length of the fish.)

5. The temperature t in that oven was over 150° C.

6. She will win the election if she gets over 35 votes. (Let V be the number of votes which will insure winning.)

290

7. He wants to buy 10 tickets and spend less than $25. (Let
    c be the cost he can spend per ticket.

8. She wants to sell 8 sweaters and get more than $200. (Let
    s be the price for one sweater.)

Lesson 7

Inequalities and Addition

Suppose your age is x and a friend's age is y. Perhaps
you are younger than this friend.

That is, perhaps $\qquad$ $x < y$

Then 10 years from now you would still be younger than your friend.

Then $\qquad$ $x + 10 < y + 10$

And 6 years ago you would still have been younger.

If $\qquad$ $x < y$

then $\qquad$ $x - 6 < y - 6$

This situation illustrates the Addition Property of Inequalities.

Addition Property
of Inequalities

> If $x < y$,
>
> then $x + a < y + a.$

If you are older than the friend, then $x > y$. a years from
now you would still be older. $x + a > y + a$. So the Addition Property of Inequalities works for < and >. And since there is a similar

291

property for equations, this idea works for $\leqslant$ and $\geqslant$ also. As a result:

> You can add the same number to both sides of an equation or inequality without affecting the solutions.

Example 1: A person owes \$5. How much must the person earn to have at least \$10 after paying the debt?

Solution: Let E be the amount the person needs to earn. Then

$$-5 + E \geqslant 10$$
$A_5:$
$$E \geqslant 15$$

Example 2: Solve for m: $m - 3 < 11$

$A_3:$
$$m < 14$$

Example 3: <u>The Triangle Inequality</u>. You have heard of the statement "The shortest distance between two points is a straight line." For a triangle with sides of lengths x, y, and z, this means that three inequalities must be satisfied.

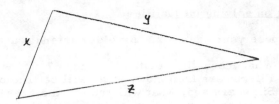

$x + y > z$ and $x + z > y$ and $z + y > x$

For example, if $x = 10$ and $y = 15$, then substituting:

$10 + 15 > z$ and $10 + z > 15$ and $z + 15 > 10$

Solve each inequality.

$25 > z$ and $z > 5$ and $z > -5$

This tells us that when two sides of a triangle are 10

and 15, the third side must be shorter than 25 and

longer then 5. (It also must be longer than -5, but

this is obvious from the start.)

The Triangle Inequality can be pictured by thinking of a

hinge connected at point H. HE is fixed. HS moves.

Hinge wide open.
Length of third
side is near 25.

Hinge almost closed.
Length of third
side is near 5.

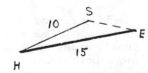

293

<u>Questions on solving inequalities</u>

1. Suppose your age is x. An older friend's age is y.

   (a) What inequality relates x and y?
   (b) Fifty years from now, what will be true of the ages?
   (c) Six years ago, what was true of the ages?
   (d) <u>a</u> years from now, what is true of the ages?

2. What is the Addition Property of Inequalities?

3-6. Apply the given operation to the sentence -4 < -1.

3. A_4 4. A_3 5. A_{-2} 6. A_{-100}

7-10. Apply the given operation to the sentence $2 + y \geq 7$.

7. A_{-2} 8. A_{-y} 9. A_{-7} 10. $A_{\frac{1}{2}}$

11-16. Solve:

11. $x + 14.2 < 20$ 12. $-9 + y > -2$

13. $-2 > 6 + a$ 14. $100 < B - 400$

15. $z - 12 > 5$ 16. $\frac{1}{2} \leq \frac{1}{2} + m$

17. Solve for m: $p + m \geq q$ 18. Solve for s: $s - t \leq 100$

19-22. (a) Solving what inequality will answer the question?
(b) Solve the inequality. (c) Answer the question.

19. A person owes $450 and has no savings. How much must the person save to pay off the debt and have at least $400 in savings?

20. Yesterday's temperature was $22°$. Today it is below $5°$. By how much did the temperature change?

21. The 1970 census gives a population of 581,000 for Phoenix, Arizona. It was estimated that there were at least a million people in the metropolitan area. How many people live in the metropolitan area but not in Phoenix itself?

22. A board is over 5 meters long. One part of length 1 m is painted. How long a piece is **not** painted?

294

Questions on the Triangle Inequality

1-3. Refer to the triangle at right.
Since the shortest distance between
two points is a straight line:

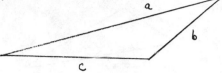

1. a + b > _____.

2. b + c > _____.

3. _____ > b.

4. What is the Triangle Inequality?

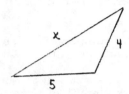

5. (a) Name three inequalities
    which involve x.
    (b) Solve each inequality
    for x.
    (c) How large can x be?
    (d) How small can x be?

6. Two metal plates are joined
    by a hinge as at right. Let
    d be the distance between
    points P and Q.

    (a) Name three inequalities
    d must satisfy.

    (b) d must be smaller than
    _____.

    (c) d must be larger than
    _____.

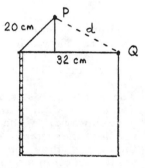

7. Why can there not be a triangle with sides of lengths 1 cm,
    2 cm, and 4 cm?

8. If two sides of a triangle have lengths 100 and 75, the third
    side can have any length between _____ and _____.

9-12. (a) Name three inequalities which the variable must satisfy. (b) Solve each inequality. (c) Use your answers to part (b) to determine the largest and smallest possible values of the variable. The drawing below is a hint for Question 11.

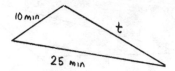

9. The Holiday Inn in Allentown, Pennsylvania, is (according to the 1973 Holiday Inn directory) 5 miles from Downtown, 10 miles from the airport, 7 miles from the Bethlehem Steel Plant, and 18 miles from Crystal Cave. Let d be the distance from the airport to the cave.

10. Refer to Question 9. Let m be the number of miles from downtown to the Bethlehem Steel Plant.

11. Betty can walk to school in 25 min. She can walk to her boyfriend's house in 10 min. Let t be the length of time for her boyfriend to walk to school. (Assume they walk at the same rate.)

12. Airline schedules allow 54 minutes to fly from Chicago to Detroit. They allow 1 hr. 42 min. to fly from Chicago to Washington. Of these times, 20 minutes are for take-off and landing. Let t be the Detroit to Washington flying time.

13. The two brightest stars in the constellation Gemini (the Twins) are Castor and Pollux. Castor is about 45 light-years away from the Earth, Pollux about 33. (a) What is the furthest these stars could be from each other? (b) What is the closest they could be? (c) These stars are in about the same direction as seen from Earth. So their distance from each other is about _____.

Inequalities and Multiplication

On any number line the numbers are in order

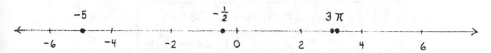

We identify some points to show the order.   Smaller numbers are at left.

$$-5 \; < \; -\tfrac{1}{2} \; < \; 3 \; < \; \pi$$

Now add 3 to each number, and graph the results

$$-2 \; < \; 2\tfrac{1}{2} \; < \; 6 \; < \; \pi + 3$$

The order is the same.   This verifies the Addition Property of Inequalities.   Even if a negative number is added to the sides of an inequality, the order will be the same.   Let us add $^-1$ to each number.

$$-3 \; < \; 1\tfrac{1}{2} \; < \; 5 \; < \; \pi + 2$$

Adding a number to both sides of an inequality is like sliding the sides of the inequality on a number line. So adding never changes the order.

> Beware: What lies ahead is
>
> simple but surprising.

Multiplying both sides of an inequality is like changing the scale on a number line. Here is the original number line.

Now we multiply each coordinate by 4.

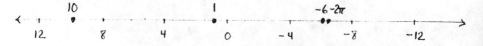

The scale is changed but not the order. Smaller numbers are still at the left. For example:

Originally: $-\frac{1}{2} < 3 < \pi$

M_4: $-2 < 12 < 4\pi$

You might now think that everything keeps the order. But recall that multiplying by a negative number changes scale and switches direction. (Think of the gear problems in Chapter 4.) For example, we multiply the coordinates by $-\frac{1}{2}$.

The order is <u>reversed</u>. Larger numbers are now at left.

Originally: $-2 \quad < \quad 12 \quad < 4\pi$

$M_{-\frac{1}{2}}$: $-2 \cdot -\frac{1}{2} > 12 \cdot -\frac{1}{2} > 4\pi \cdot -\frac{1}{2}$

$1 \quad > \quad -6 \quad > -2\pi$

These examples illustrate the Multiplication Property of Inequalities.

Multiplication Properties of Inequality:	Suppose $x < y$. Then if a is positive, $ax < ay$. But if a is negative, $ax > ay$.

Example 1: Solve $\frac{x}{14} \leqslant \frac{3}{5}$

M_{14}: $x \leqslant \frac{42}{5}$ Multiplying both sides by 14 keeps the order.

Example 2: Solve $-4A > 12$

$M_{-\frac{1}{4}}$: $A < -3$ Multiplying both sides by $-\frac{1}{4}$ reverses the order.

To check that <u>A < ⁻3</u> is the correct answer to <u>⁻4A >12</u>, you must do two things. First, check the number ⁻3 by substituting it in the original sentence. It should make both sides equal.

$-4 \cdot -3 \overset{?}{=} 12$ It makes them equal.

299

Second, check the $<$ inequality sign by choosing a number which works in $\underline{A < -3}$. (You might choose -100.) This number should also work in the original sentence.

$$-4 \cdot -100 \overset{?}{>} 12$$

$$400 > 12 \qquad \text{It works.}$$

You should try never to check a problem by doing it over. If there is an error, most people make the same error a second time.

Questions covering the reading

1-4. Here are some numbers in order. Perform the given operation on the numbers. Tell whether the order is kept or reversed.

$$-8 < -3 < 5 < 10\tfrac{1}{2}$$

1.   Multiply each number by 2.

2.   Multiply each number by -6.

3.   Multiply each number by -2.

4.   Multiply each number by .70.

5-8. Here is a sentence. What sentence results from applying each operation?

$$100A \geqslant 400$$

5. $M_{\frac{1}{100}}$ 6. A_{-100} 7. M_{-2} 8. M_{-100}

9.   What are the Multiplication Properties of Inequality?

10. A student solved $-2x < -6$
and got $x < 3$

    (a) What did the student do wrong?
    (b) How can the "check" show the error?

11. What are the two steps in checking solutions to inequalities?

12-13. Here is a problem and an answer. <u>Check</u> the work.

12. Problem: $-6x < \frac{2}{3}$

    Answer: $x > -\frac{1}{9}$

13. Problem: $12 \geqslant 3y$

    Answer: $4 \leqslant y$

14. <u>Multiple Choice</u>. In which of these operations should the direction of an inequality be reversed?

    (a) $M_{\frac{1}{10}}$ (b) $A_{\frac{1}{10}}$ (c) M_{-10} (d) A_{-10}

15-24. Solve each sentence and check the solutions.

15. $12x > 60$

16. $-50W \geqslant 10$

17. $3 > \frac{x}{3}$

18. $\frac{A}{9} < \frac{2}{3}$

19. $-3y > -9$

20. $-5B > 30$

21. $200 \leqslant -\frac{1}{10} C$

22. $\frac{1}{5} \leqslant \frac{2}{5}t$

23. $2 < -2m$

24. $20n < \frac{4}{5}$

25. Why should you not check an answer by doing the problem over?

<u>Questions applying the reading</u>

1-4. If A is the area of the parallelogram drawn below, then $A \leqslant \ell w$.

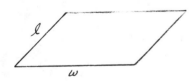

1. For an area of 50 sq cm, if w is 5 cm long, then what must ℓ be?

2. What are the possible values of w if $\ell = 40$ and $A = 30$?

3. Solve the formula for ℓ.   4. Solve the formula for w.

301

5. The Multiplication Properties of Inequalities ignore the possibility that you might wish to multiply both sides of $x < y$ by 0. What happens then?

6. Use the formula $d = rt$. If you wish to travel over 100 km in 3 hr., what must be your mean rate?

7. Suppose pads of paper cost 50¢. If you have less than $7, what are the possible numbers of pads of paper you can buy?

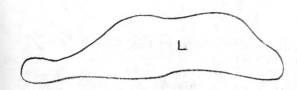

8-10. The region L is the picture of a small lake. Let B be the length of the shore of L. Let A be the area of L. A and B are related by the formula

$$4 \pi A \leqslant B \cdot B$$

(This formula tells the largest area you can have with a given perimeter.)

C 8. What is the largest possible area of L if the length of shore is 1 mile? (Use the approximation 3.14 for π.)

C 9. A farmer buys 2000 m of fence to surround his cattle. (a) What is the area which can be bounded by the fence if it is arranged in a square? (b) What is the largest area which can be bounded by the fence?

10. Draw a lake in which B is large but A is very small.

11-14. Remember that $-x = -1 \cdot x$. Solve each sentence.

11. $-x > 2$

12. $-y < -5$

13. $\frac{1}{7} \geqslant -z$

14. $0 \leqslant -w$

302

Lesson 9 (Optional)

More Problems to Solve

You have now learned how to solve sentences involving any one of the four fundamental operations. Until now, all questions in a given lesson have followed the idea of that lesson. So you knew which operation to apply.

In real life, problems do not follow lessons. Faced with a problem, you must figure out what lesson to apply.

For the questions below:

    (a) Give an equation or inequality which

        could help answer the question.

    (b) Solve the equation or inequality.

    (c) Answer the question.

If you have trouble answering part (a), ask yourself: What operation seems to be involved here? Is there a model which applies to this situation?

If you still have trouble, ask yourself: Could I answer the problem if the numbers were simpler? Then try to do that. See if you can find the pattern using simpler numbers.

If you still have trouble, look at the answer to part (a) upside down on page 306.

Around the Home

1. A recipe for marshmallow treats (on the box of Kellogg's Rice Krispies) uses the following ingredients and yields 24 squares.

    1/4 cup butter
    40 regular marshmallows
    5 cups rice krispies

    How many squares could be made if you had butter and rice krispies but only 25 regular marshmallows?

2. All members of the Jones family like 4 desserts, 3 vegetables, and 5 appetizers. How many main courses must the family like in order to be able to have a different meal each day of the year?

3. A wall is 12' 3" long and you have a couch which is 84" wide. If you want at least 5" between the couch and a table and 1' on each outer side of this furniture, how wide a table can you buy?

4. If you like beef done "medium," the Better Homes and Gardens Cookbook recommends that you cook a standing roast for about 45 minutes at 325° for each kg of weight. How large a roast can be cooked in 2 1/2 hours?

5. A ceiling is 2.6 meters high and a door is 2.1 meters high. How much space is there between the top of the door and the ceiling?

Money Problems

1. A store ad says "Save over $20 on every chair. All prices reduced 30%." What was the original price of a chair?

2. Suppose you have $110.37 in a checking account. The bank does not charge for checks if $50 or more is in the account. For how much can you make a check and not be charged?

C 3. Suppose you want an income of $10,000 a year and you feel you can earn 8% on an investment. How much must you invest? (Use the formula $I = prt$ but let $t = 1$.)

4. Suppose you have H dollars, want to buy something which
    costs C dollars, and need N dollars. Solve for N in terms
    of H and C.

5. On the day this problem is being written, the author bought
    an 8.5 oz bag of Cheetos for 69¢. There is also a 49¢ bag.
    At the rate of the 8.5 oz. bag, how many oz. of Cheetos
    should be in the 49¢ bag?

Physical Situations

1. The Earth and Venus revolve around the Sun in nearly cir-
    cular orbits. The Earth is about 150 million kilometers from
    the Sun. Venus is about 108 million km from the sun. Let
    d be the distance between the Earth and Venus. What can
    be said about d?

2. A storefront is 20 m wide, has 2 floors, and is advertised
    as containing 2500 sq m of floor space. How deep is the
    store?

3. When the temperature rose from $-4°$ F to $45°$ F in 2 minutes
    in Spearfish, South Dakota (January 22, 1943, as cited in the
    Guiness Book of World Records), by how much did it rise?

4. The record parachute escape is reported to have been made
    by two British Royal Air Force officers in 1958. Their
    plane exploded at a height of 17 kilometers and they fell to
    3 kilometers before their parachute opened. How many
    kilometers did they fall without the aid of a parachute?

5. If it takes you 3 minutes to address 7 Christmas cards, how
    long will it take you (at the same rate) to address 30 cards?

C   1.   Recall that the "earned run average" of a pitcher in baseball
satisfies the following formula:

$$\frac{\text{earned run average}}{9} = \frac{\text{earned runs allowed}}{\text{no. of innings pitched}}$$

The <u>Official Major League Baseball Record Book</u> gives the
earned run average and number of innings pitched for every
player in Baseball's Hall of Fame. But this book does not
tell you how many earned runs were allowed.

Babe Ruth (who broke in as a pitcher) pitched 1221 innings
with a lifetime earned run average of 2.28. How many
earned runs did he allow?

C   2.   A Boeing 747 "Jumbo Jet" has a length of about 70.5 meters
and a wingspan of about 59.7 meters. If a model of this
plane is 50 cm long, how wide is the model?

3.   A skating meet has 6 judges' scores which count. If a skater
needs over 33 points in order to win a medal, how much on
the average is needed from each judge?

4.   A hockey team feels it needs at least 60 points to make the
playoffs. Now it has 23 points. How many points are needed?

5.   How many 13¢ stamps can you buy for $10?

Answers to part (a): Around the Home: 1. $\frac{40}{25} = \frac{24}{s}$ 2. $4 \cdot 3 \cdot 5 \cdot m \geqslant 366$ 3. $12 + 84 + 5 + W + 12 = 111$ 4. $\frac{45}{1} = \frac{x}{150}$ 5. $2.1 + s = 2.6$ Money: 1. $30p > 20$ 2. $110.37 = 110.37 + c = 50.00$ 3. $.081 = 10000$ 4. $H + N = C$ 5. $\frac{69}{49} = \frac{8.5}{C}$ Physical: 1. $d + 108 \geqslant 150$, $108 \geqslant 150 + d$ 2. $20 \cdot d = 2500$ 3. $-4 + r = 45$ 4. $17 - f = 3$ 5. $\frac{1}{3} = \frac{7}{30}$ Sports and Hobbies: 1. $\frac{E}{2.28} = \frac{1221}{9}$ 2. $\frac{70.5}{59.7} = \frac{50}{W}$ 3. $6s > 33$ 4. $23 + N \geqslant 60$ 5. $13n \leqslant 10$

306

Chapter Summary

Many real situations can be translated into mathematical sentences. Finding the values of a variable which make a sentence true is called solving the sentence. There are standard procedures for solving many sentences. These procedures are called _algorithms_. All algorithms for sentence-solving are based upon properties of equations and inequalities.

The following chart shows how the ideas of this chapter are related. You must know each idea in order to understand the ones below it.

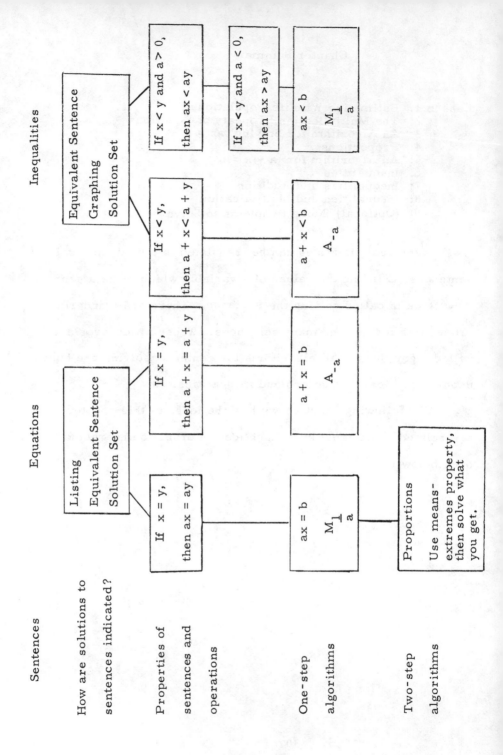

Sentences

Equations

Inequalities

How are solutions to
sentences indicated?

Listing
Equivalent Sentence
Solution Set

Equivalent Sentence
Graphing
Solution Set

Properties of
sentences and
operations

If $x = y$,
then $ax = ay$

If $x = y$,
then $a + x = a + y$

If $x < y$,
then $a + x < a + y$

If $x < y$ and $a > 0$,
then $ax < ay$

If $x < y$ and $a < 0$,
then $ax > ay$

One-step
algorithms

$ax = b$

$M_{\frac{1}{a}}$

$a + x = b$

A_{-a}

$a + x < b$

A_{-a}

$ax < b$

$M_{\frac{1}{a}}$

Two-step
algorithms

Proportions

Use means-
extremes property,
then solve what
you get.

308

CHAPTER 7

LINEAR EXPRESSIONS AND DISTRIBUTIVITY, Part I

Lesson 1

Situations Leading to Linear Expressions

$\underline{1.03x + 2}$ $\underline{9(2 - 4y)}$ $\underline{3A + 4B + C}$ $-\frac{1}{5}M$ $\frac{x + 2}{3}$

Underlined are linear expressions. In linear expressions, the variables may be added or subtracted. Variables may be added to, multiplied by, divided by, or subtracted from real numbers. In a linear expression two variables may not be multiplied or divided, nor may a real number be divided by a variable. Here are some expressions which are not linear.

$$xy \qquad \frac{a + 3}{b - 2} \qquad v(w + z) \qquad \frac{3}{t}$$

(Linear expressions have the name "linear" because they often lead to graphs which are lines. You will study these graphs in Chapter 9.)

Many real situations lead to patterns which are described by linear expressions. Here are some examples.

1.   Weight of an orange crate (a constant gain example)

A crate of oranges might weigh about 6 kg when empty. A typical orange weighs .2 kg. How much will the crate weigh when

309

it is filled with n oranges? To answer this question, we make
a chart.

no. of oranges	weight of crate (in kg)
0	6
1	6 + .2
2	6 + 2 · .2
3	6 + 3 · .2
4	6 + 4 · .2
.	.
.	.
.	.

The chart does not end. But the entire chart can be described
in one line.

$$n \qquad\qquad 6 + n \cdot .2$$

That is, a crate with n oranges weighs 6 + .2n kg.

2.   Population (a constant loss example)

Suppose a town of 25,000 people is losing 500 people a year
in population. How many people will be in the town y years from
now?

Years from now	population
0	25,000
1	25,000 - 500
2	25,000 - 2 · 500
3	25,000 - 3 · 500
4	25,000 - 4 · 500
.	.
.	.
.	.
y	25,000 - y · 500

There would be 25,000 - 500y people in the town y years from now.

310

3.   Total Sales (a sum of products example)

For a live concert, not all seats cost the same. Suppose E

seats are sold at $6.50, M seats are sold at $5.50, and C seats

are sold at $4.50. What will be the total income?

Answer: Income from the expensive seats will be 6.50E.

Income from the moderate seats will be 5.50M.

Income from the cheapest seats will be   4.50C.

Total income: 6.50E + 5.50M + 4.50C

Questions covering the reading

1.   Give an example of a linear expression.

2.   Give an example of an expression which is not a linear expression.

3.   How do "linear expressions" get their name?

4.   Refer to the orange crate example on page 309. Give the weight of the orange crate when:

(a) empty.   (b) filled with 1 orange.
(c) filled with 5 oranges.   (d) filled with n oranges.
(e) filled with 120 oranges.

5.   Suppose a crate weighs 10 kg when empty. You fill it with grapefruit. Each dozen grapefruit weighs about 5 kg. Finish this table:

no. of grapefruit	weight of crate
0	(a) _____
1 dozen	(b) _____
2 dozen	(c) _____
3 dozen	(d) _____
.	
.	
.	
n dozen	(e) _____

311

6. A town is gaining population at a rate of about 200 per year. There are now 10,000 people in the town. Finish this table:

years from now	population of town
0	(a) _____
1	(b) _____
2	(c) _____
3	(d) _____
.	
.	
.	
n	(e) _____

7. (Review) Evaluate the expression $15 + 10n$ when

(a) $n = 0$ (b) $n = 1$ (c) $n = -1$ (d) $n = 6$

8. (Review) Find the value of $25000 - y \cdot 500$ when

(a) $y = 10$ (b) $y = 6$ (c) $y = -2$ (d) $y = 50$

9-10. Answer the question. Then look at Example 3, p. 311, and tell what your answer means.

C 9. Find the value of $6.50E + 5.50M + 4.50C$ when $E = 200$, $M = 500$, and $C = 250$.

C 10. Evaluate $6.50E + 5.50M + 4.50C$ when $E = 500$, $M = 450$, and $C = 0$.

11-12. Give the total income you would receive from selling:

11. m tickets at \$3, n tickets at \$2, and t tickets at \$1.

12. 800 tickets at \$x and 500 tickets at \$y.

13-14. How much would it cost you to buy:

13. c cookies at 2¢ apiece and 80¢ for some milk.

14. \$5 admission to an amusement park and r rides at 50¢ each.

312

Questions extending the reading

1. An envelope weighs about 5 grams. A sheet of letter paper
   weighs about 3 grams.
   (a) Make a chart showing how much a mailed letter will
       weigh if it contains 1, 2, 3, 4, and 5 pages of inserts.
   (b) What will be the total weight of the mailed letter if it
       has p pages of inserts?

2. The Williams family wishes to save money. They decide
   to start a special account with $50 and add $15 each week.
   (a) Make a chart showing how much the Williamses will
       have after 1, 2, 3, 4, 5, ... weeks.
   (b) How much will they have saved after n weeks?

3. Suppose you have a 15-gallon aquarium which weighs 28 lbs.
   when empty. A gallon of water weighs about 8 1/3 lbs.
   (a) Make a chart to show how much the aquarium will weigh
       when filled with 1, 2, 3, 4, 5, ... gallons of water.
   (b) How much will the aquarium weigh when filled with W
       gallons of water?

4. An obese man now weighs about 154 kg. He hears of a special
   hospital in Switzerland in which people lose as much as 2.5 kg
   a week. He thinks he can lose weight at this rate.
   (a) Make a chart to show how much he would weigh after 1, 2,
       3, 4, 5, ... weeks.
   (b) How much would he weigh after t weeks of treatment?

5. A typical long distance telephone rate is $0.65 for the first 3
   minutes and $0.20 a minute for each additional minute. Here
   is a chart of times and rates.

minutes	cost
3	.65
4	.65 + .20
5	.65 + .20 · 2
6	.65 + .20 · 3
7	.65 + .20 · 4
.	.
.	.
.	.

   (a) How much does a 5-minute call cost?
   (b) How much does a 30-minute call cost?
   (c) How much does a call cost if it is t minutes long?

6.   Taxicab rates. A taxicab charges 50¢ for the first $\frac{2}{5}$ mile and 10¢ each succeeding $\frac{1}{5}$ mile.

(a) Make a chart to show how much it costs to go $\frac{2}{5}$, $\frac{3}{5}$, $\frac{4}{5}$, ..., miles.

(b) Find a pattern to show how much it costs to go $\frac{n}{5}$ miles.

(c) If you live 10 miles from an airport, how much will it cost you to take this cab there?

7.   Suppose you have a 60-liter aquarium which weighs 12 kg when empty. A liter of water weighs exactly 1 kg.

(a) Make a chart to show how much the aquarium will weigh when filled with 1, 2, 3, 4, 5, ... liters of water.

(b) How much will the aquarium weigh when filled with W liters of water?

Lesson 2

Situations Leading to ax + b = c or ax + b < c

The sentences in the title of this lesson involve only linear expressions. So they are called linear sentences. Situations which lead to linear expressions naturally lead to linear sentences.

Situation 1: A crate weighs 6 kg when empty. An orange weighs about .2 kg. With n oranges, the crate's total weight (found in Lesson 1) is

6 + .2n

314

Question 1: Suppose the crate can weigh 50 kg without break-

ing. How many oranges can it hold?

Answer: We want to know--for what value of n

is the weight 50 kg? So we solve

$$6 + .2n = 50$$

Question 2: The crate must weigh at least 45 kg to be econom-

ical. How many oranges should be put in the crate?

Answer: Solve $6 + .2n \geqslant 45$

Every situation can have many questions asked about it. Here
is a situation involving subtraction.

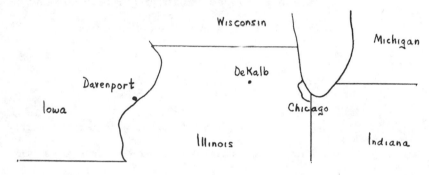

Situation 2: A line of storms is at Davenport, Iowa, and mov-

ing east towards Chicago at 20 mph. Davenport

is about 150 miles west of Chicago. Moving at

this rate the storms will be 150 - 20h miles

away in h hours.

Question 1:   When will the storms hit DeKalb, 60 miles west
                 of Chicago?

                 Answer: Solve $150 - 20h = 60$.

Question 2:   When were the storms 200 miles west of Chicago?

                 Answer: Solve $150 - 20h = 200$.

        The next situation requires two variables.

Situation 3:   One ounce of hamburger meat has about 80 cal-
                 ories, a bun has about 200 calories, and 1
                 "french fry" has about 14 calories. Then if
                 you eat h ounces of hamburger on a bun and
                 f french fries, you will have taken in

                 80h   +   200   +   14f   calories.
                  ↑   ↑   ↑
                 from   from   from
              hamburger   bun   fries

Question 1:   Suppose you order a 4 oz. hamburger (a quarter
                 pound). How many french fries can you eat and
                 still be under 600 calories?

                 Answer: $h = 4$ here. So substitute 4 for h.
                 Then solve

                         $320 + 200 + 14f < 600$

Question 2:   In the situation of Question 1, how many french
                 fries can you eat if you throw away the bun?

                 Answer: Solve $320 + 14f < 600$

                                316

Question 3:   If you eat 25 french fries, how big a hamburger

can you order and get 1000 calories for this

food?

<u>Answer</u>: f = 25 here. You don't know h. So

solve

$$80h + 200 + 14 \cdot 25 = 1000$$

Of course the big issue is--how can these sentences be solved?
That is the topic of Lesson 3.

<u>Special note to students</u>: In Lessons 3 and 4, you will be asked to
solve some of the answers you get below.   So you may find it helpful
to write down your answers to these questions in a place where you
can refer to them.

<u>Questions covering the reading</u>

1-4. A given crate weighs 10 kg when empty. A grapefruit weighs
about 1/2 kg. Let n be the number of grapefruit in the crate.
<u>What sentence could be solved</u> to answer each question?

1.   How many grapefruit can be put in the crate and still keep the
      total weight under 40 kg?

2.   How many grapefruit can be put in the crate so that the weight
      is over 35 kg?

3.   How many grapefruit can be put in the crate so that the weight
      is over 50 kg?

4.   How many grapefruit can be put in the crate so that the weight
      is under 25 kg?

5-8. Refer to Situation 3, page 316. <u>What sentence could be solved</u>
to answer each question?   (Assume that every hamburger comes
with a bun.)

5.   You order a 3 oz. hamburger. How many french fries can you
      eat and still be under 600 calories?

317

6.   You order a 5 oz. hamburger. How many french fries can you eat if you need at least 800 calories?

7.   If you eat no french fries, how much hamburger could you eat if you wanted at least 650 calories?

8.   If you eat 30 french fries, how much hamburger could you eat if you wanted no more than 400 calories?

9-12.   Refer to Situation 2, page 315. <u>What sentence could be solved to answer each question</u>? Assume the storm line moves 20 mph to the east.

9.   When will the storms hit Aurora, 35 miles west of Chicago?

10.   When will the storms hit South Bend, 45 miles <u>east</u> of Chicago?

11.   When were the storms 200 miles west of Chicago?

12.   When were the storms 150 miles west of Chicago?

<u>Questions applying the reading</u>

1-3.   An elevator has a capacity of 1 metric ton. (1 metric ton = 1000 kg.) Adult men average about 80 kg in weight; adult women average about 55 kg in weight. <u>What sentence could be solved</u> to answer each question?

1.   If 3 women are on the elevator, how many men could get on and still keep the weight under capacity.

2.   If 4 men are on the elevator, how many women could get on and still keep the weight under capacity?

3.   How many men could get on the elevator with a 150 kg piano?

4-6.   A person pays $500 down and $150 a month for a car. <u>What sentence could be solved</u> to answer each question?

4.   In how many months will the person have paid $2500 total?

5.   In how many months will the person have paid $3500 total?

6.   In how many months will the person have paid $4000 total?

7-9. A person goes into a record store with $50.39. All albums the person wants are on sale for $4.79. <u>What sentence could be solved</u> to answer each question?

7.   How many records can the person buy and still have at least $20.00 left?

8.   How many records can the person buy and still have at least $10.00 left?

9.   How many records can the person buy?

10-12. A homemaker saves coupons from cake-mix boxes. Each box gives 3 coupons and two boxes are used in a week. 128 coupons have been saved so far. <u>What sentence could be solved</u> to answer each question?

10.   In how many weeks will 250 coupons be saved?

11.   In how many weeks will 312 coupons be saved?

12.   When were 100 coupons saved?

Lesson 3

<u>An Algorithm for Solving $ax + b = c$</u>

From asking how many oranges an orange crate could hold, the following equation can arise.

$$6 + .2n = 50$$

This equation can be solved in two steps.

Step 1: A_{-6}: $-6 + 6 + .2n = -6 + 50$

            Simplify: $.2n = 44$

Now the equation is one you can solve.

319

Step 2: $M_{\frac{1}{.2}}$: $n = \dfrac{44}{.2}$

        Simplify: $n = 220$

The crate could hold 220 oranges.

        Question 3 in Situation 3, page 316, led to the next equation.

$$80h + 200 + 14 \cdot 25 = 1000$$

Here h is the number of ounces of hamburger you can eat if you
want 25 french fries and gain 1000 calories. To solve, first
simplify.

$$80h + 200 + 350 = 1000$$
$$80h + 550 = 1000$$

Now it again takes two steps.

Step 1: A_{-550}: $80h = 450$

You have solved this kind before.

Step 2: $M_{\frac{1}{80}}$: $h = \dfrac{450}{80}$

        Simplify: $h = 5.625$

You can eat a little over $5\frac{1}{2}$ ounces of hamburger meat.

        The previous examples suggest that every sentence of the
form

$$ax + b = c$$

320

can be solved in two steps. First, add -b to each side.

Step 1: A₋ᵦ: $ax = c + -b$

Now the equation is like those you know how to solve.

Step 2: M₁: $x = \dfrac{c + -b}{a}$
$\overline{}$
a

Algorithm 3: To solve $ax + b = c$ for x, first add $-b$

to each side. Then solve the resulting

equation.

Example 1: Solve: $3n + 14 = -10$

Solution: A₋₁₄: $3n = -24$

              M₁: $n = -8$
              ‾3

Check by substitution:

$3 \cdot -8 + 14 \stackrel{?}{=} -10$

$-24 + 14 \stackrel{?}{=} -10$

Yes, it checks.

Example 2:   A line of storms is at Davenport, Iowa, and mov-

ing east towards Chicago at 20 mph. Davenport

is about 150 miles west of Chicago. At this rate,

when were the storms 200 miles west of Chicago?

(This is Question 2 of Situation 2 on page 316. See

the drawing, page 315.)

321

Solution: In h hours, the storms are 150 - 20h miles

away.   So you need to solve

$$150 - 20h = 200$$

A_{-150}: $-20h = 50$

$M_{\frac{-1}{20}}$: $h = -\dfrac{50}{20}$

$$h = -2.5$$

That is, $2\frac{1}{2}$ hours <u>ago</u> the storms were 200 miles

west of Chicago (If h were positive, h would

stand for time in the future.)

Because equation-solving can be tricky, many students find

it helpful to show all work.   This is done in the next example.

Example 3: Solve $\dfrac{2}{3} = \dfrac{3}{4} x - \dfrac{2}{5}$

Solution: Step 1: $A_{\frac{2}{5}}$: $\dfrac{2}{3} + \dfrac{2}{5} = \dfrac{3}{4} x - \dfrac{2}{5} + \dfrac{2}{5}$

Simplify: $\dfrac{10}{15} + \dfrac{6}{15} = \dfrac{3}{4} x$

$$\dfrac{16}{15} = \dfrac{3}{4} x$$

Step 2: $M_{\frac{4}{3}}$: $\dfrac{4}{3} \cdot \dfrac{16}{15} = \dfrac{4}{3} \cdot \dfrac{3}{4} x$

Simplify: $\dfrac{64}{45} = x$

Check by Substitution:

$$\frac{2}{3} \overset{?}{=} \frac{3}{4} \cdot \frac{64}{45} - \frac{2}{5}$$

$$\frac{2}{3} \overset{?}{=} \frac{16}{15} - \frac{2}{5}$$

$$\frac{10}{15} \overset{?}{=} \frac{16}{15} - \frac{6}{15} \qquad \text{Yes, it checks.}$$

Questions covering the reading

1-10. (a) To help solve the given equation, what number should be added to each side? (b) Solve the equation.

1. $2v + 3 = {}^-4$

2. $6 + 14m = 90$

3. ${}^-2 + 7y = 26$

4. ${}^-47 = {}^-7 - 20x$

5. $8 = 4x + 9$

6. ${}^-4A - 119 = 211$

7. $6 - 2v = {}^-40$

8. $\frac{2}{3}x + 10 = 4$

9. $1000t - 300 = 500$

10. $\frac{1}{2} = 4 - \frac{3}{2}x$

11.   To solve $ax + b = c$ for x, what can you do?

12.   To solve $5y - t = q$ for y, what can you do?

13-16. Refer to the situation of Example 2, p. 321. (a) Give an equation which can be solved to answer each question. (b) Solve the equation. (c) Answer the question.

13.   When were the storms 200 miles west of Chicago?

14.   When were the storms 150 miles west of Chicago?

15.   When will the storms hit Aurora, 35 miles west of Chicago?

16.   When will the storms hit South Bend, 45 miles _east_ of Chicago?

17-20. A crate weighs 6 kg when empty and can hold 40 kg without breaking. (a) Give an equation which can be solved to answer each question. (b) Solve the equation. (c) Answer the question.

17.   How many oranges weighing .2 kg apiece can the crate hold?

18.   How many lemons weighing .15 kg apiece can the crate hold?

323

19. How many grapefruit weighing .5 kg apiece can the crate hold?

20. How many canteloupe weighing .8 kg apiece can the crate hold?

21-22. Solve and check.

21. $4t + 78 = 150$

22. $11 = 17 + 2u$

23. $9x - 20 = 142$

24. $11y + 6 = -5$

25. $8z + 2 = 5$

26. $-3W - 18 = 12$

27. $2 - 5A = 12$

28. $7 = 6 + 2B$

29. $\frac{1}{2} - \frac{C}{3} = 2$

30. $-30 = 100 - 70D$

31. $-2 = \frac{E}{5} + 6$

32. $7.2 + .3G = 3.9$

Questions applying the reading

1-2. Suppose you begin a Christmas account with $80. These accounts often do not give any interest.

1. If you add $25 per week, how many weeks will it take you to have $250?

2. If you want to save $300 by December 1st and can add $15 a week, how many weeks before December 1st should you start this account?

3. An automatic freight elevator has a capacity of 2000 kg. How many pianos weighing 150 kg can it lift at one time if a worker weighing 80 kg must ride with the pianos?

4. Answer Question 3 if the capacity of the elevator is 2500 kg.

5-6. One way to earn spare change is to pick up aluminum cans and soft drink bottles which are litter in some places. You can often earn 1¢ for each can and 5¢ for a bottle.

5. If you want $10.00 and find 630 cans, how many bottles will you need?

6. If you want $10.00 and find c cans, how many bottles will you need?

7-8. To find the length L of a belt going around two wheels, machinists use the formula L = 3.26 (r + R) + 2D, where R = radius of large wheel, r = radius of small wheel, D = distance between centers of wheels.

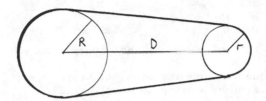

7. Suppose you have a belt of length 150 cm and two wheels with radii 18 cm and 10 cm. How far apart should the centers of the wheels be?

8. Answer Question 7 if each wheel has radius 10 cm.

9-12. Answer each of questions 9-12 on page 319 by writing an equation and solving it.

13-14. An envelope weighs about 5 grams. A sheet of letter paper weighs about 3 grams. First class postal rates as of 1976 were 13¢ for the first ounce, and an ounce is about 28.35 grams.

13. How many sheets could you put in an envelope and still pay only 13¢ for mailing?

14. Answer Question 13 if you use a heavier envelope weighing 7 g and lighter paper weighing 2.5 g.

15-16. Answer each of Questions 5 and 6 on page 318 by writing an equation and solving it.

17-18. An obese woman goes to the same special hospital in Switzerland the man went to on page 313. This is the hospital where people lose as much as 2.5 kg a week. She now weighs 130 kg.

17. At least how long will it take for her to get down to 110 kg?

18. How much time will it take her to get down to 105 kg?

19-22. Simplify each sentence. Then solve.

19. $22 = 2 + 2Z + 2$ 20. $8m + 3 + 5 = 0$

21. $11 - 2x - 6 = 5$ 22. $11 - 6y = 35 + 6$

Questions for discussion

1. The process of building up ax + b can be described in the
    following manner.

Start:	x
Multiply by a:	ax
Add b:	ax + b

    How is this related to solving the equation ax + b = c ?

2. Make up a question like the previous Questions 1-18 which
    can be answered by solving an equation.

Lesson 4

An Algorithm for Solving ax + b < c

The algorithm for solving ax + b < c is <u>identical</u> to the one
which helps solve ax + b = c.

Algorithm 4: To solve ax + b < c for x, first add -b to

each side.   Then solve the resulting inequality.

Example 1:   (Question 2 on page 315.) An orange crate weighs

6 + .2n kilograms when filled with n oranges. To

be economical, it must weigh at least 45 kg. How

many oranges should be put in the crate? To ans-

wer this solve:

$$6 + .2n \geq 45$$

$$6 + .2n \geqslant 45$$

Solution: A_{-6}: $.2n \geqslant 39$

$M_{\frac{1}{.2}}$: $n \geqslant \dfrac{39}{.2}$

Simplify: $n \geqslant 195$

So at least 195 oranges should be put in the crate.

The steps in the solution of Example 1 are the same steps as in the solution of $6 + .2n = 50$ shown on page 319. Only the arithmetic is different.

Example 2:   If you order a 4 oz. hamburger and throw away the bun, how many french fries can you eat and still be under 600 calories?

Solution: Using the work done on page 316, let f be the number of french fries eaten. Then

$$320 + 14f < 600$$

A_{-320}: $14f < 280$

$M_{\frac{1}{14}}$: $f < 20$

You must eat less than 20 french fries.

Example 3:   Solve: $11 < 2 - \frac{1}{3}t$

Solution: A_{-2}: $9 < -\frac{1}{3}t$

M_{-3}: $-27 > t$ Multiplying by a negative number switches the order.

327

Except for possibly complicated arithmetic, no problem of this type is more difficult than Example 3.

Questions covering the reading

1. True or False: In solving sentences (a) and (b) below, you could use the same first step.

    (a) $4 - 5t < 3$ (b) $4 - 5t = 3$

2. According to the algorithm of this lesson, what is the first step in solving sentence (a) above? Solve the sentence.

3-6. Give the first step in solving each sentence. Then solve.

3. $3x + 4 < -5$ 4. $5 + 13y > 70$

5. $-2 + 14z \geqslant 33$ 6. $-48 < -8 - 20A$

7-20. Solve.

7. $8 \leqslant 4B + 10$ 8. $-3C - 77 \geqslant -98$

9. $5 - 3D > -40$ 10. $\frac{2}{3} \leqslant 10 - \frac{h}{3}$

11. $-10x + 18 \geqslant 25$ 12. $133 > 9x - 11$

13. $.03y + .12 > .27$ 14. $3t + 10 < 10$

15. $-3A - 25 \leqslant 35$ 16. $10 - 9B \geqslant 100$

17. $\frac{3}{10} > \frac{2}{5} - 3C$ 18. $1000 < 50D + 200$

19. $8 - 8u < 8$ 20. $5 + 4t < -9$

21-23. One ounce of hamburger meat has about 80 calories, a bun has about 200 calories, and 1 french fry has about 14 calories. (a) Write an inequality which will help answer the given question. (b) Solve the inequality. (c) Answer the question.

21. Suppose you order a 4 oz. hamburger (a quarter-pound) with a bun. How many french fries can you eat and still be under 600 calories?

22. You get a 3 oz. hamburger with a bun. How many french fries must you eat to have at least 800 calories?

23. If you eat no french fries, how much hamburger must you have with a bun if you want 650 calories?

24. To solve $m < np + q$ for p, what is an appropriate first step?

25. Review. What are the two steps to checking your answer to an inequality?

Questions testing understanding of the reading

1-4. Answer the question asked in Questions 1-4 on page 317.

5-6. Six people are given tickets to a football game and $8 for lunch there. Hot dogs go for 65¢ and drinks for 25¢. They buy h hot dogs and d drinks.

5. If $h = 9$, what is the largest value that d could have?

6. If each person has one hot dog, how many cola drinks could be bought?

7-8. Answer the question asked in Questions 7-8 on page 319.

9-10. It takes a carpenter 8 hours to make a chair, 12 hours to make a table, and 20 hours to make a bookcase. Let c, t, and b be the number of chairs, tables, and bookcases that are made.

9. (a) For 25 working days of 8 hours each, what sentence describes how c, t, and b are related?

   (b) This carpenter receives orders for 2 tables, 8 chairs, and a bookcase. After these are done, how many chairs would the carpenter be able to make in the 25 days?

10. (a) In 30 working days of 9 hours each, how are c, t, and be related?

    (b) If the carpenter made 4 chairs, a table, and a bookcase, how many more tables could be made in these 30 days?

329

11. A card printer charges $5.00 to set up each job and then $.75 for each 100 cards printed.
    (a) How much would it cost you to print n hundred cards?
    (b) How many cards could you print for under $25?

12. An average 45 RPM record cut takes $2\frac{1}{2}$ minutes, one side of an LP takes about 20 minutes. If you own 30 45 RPM records, how many LP's would you need to have to make sure of being able to play for at least 4 hours?

13-18. Simplify. Then solve.

13. $3 + 2H + 5 < 30$

14. $6 - 2n - 5 > 3$

15. $23 - 3 - 7y \leqslant -8$

16. $4x + 9 - 20 \geqslant 5 - 11$

17. $-12 > -2z + 8 - 8$

18. $0 < 3t - 9 - 9$

Lesson 5

The Distributive Property

Situation 1:   Monday you buy 7 candy bars at 15¢ each.

Tuesday you buy 3 candy bars at 15¢ each.

How much have you spent?

Answer: You have bought 10 candy bars at 15¢ each, so you have spent $1.50.

This situation shows that $7 \cdot .15$ + $3 \cdot .15$ = $10 \cdot .15$

↑	↑	↑
spent	spent	Total
Monday	Tuesday	spent

Check: 1.05 + $.45$ = 1.50

330

The same idea holds regardless of the cost of a candy bar. If the cost of one bar is c, then:

$$\underset{\substack{\uparrow \\ \text{spent} \\ \text{Monday}}}{7c} \quad + \quad \underset{\substack{\uparrow \\ \text{spent} \\ \text{Tuesday}}}{3c} \quad = \quad \underset{\substack{\uparrow \\ \text{Total} \\ \text{spent}}}{10c}$$

The equation 7c + 3c = 10c is true for <u>any</u> value of c. You could think of c as negative (money paid out is often negative). Suppose each bar cost 9¢. Check that this is true:

$$7 \cdot {}^-9 \quad + \quad 3 \cdot {}^-9 \quad = 10 \cdot {}^-9$$

Why does this work? Think of paying for the bars one at a time. The repeated addition leads to multiplication.

$$= \overbrace{c + c + c + c + c + c + c}^{7c} + \overbrace{c + c + c}^{+ \quad 3c} = 10c$$

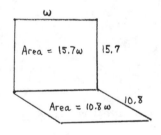

Situation 2: You wish to cover the back and one shelf of a cabinet. The width is w and other dimensions are as shown at left. How much area needs to be covered?

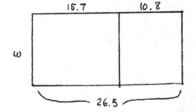

Answer: It is easiest to think of laying the pieces flat. The total length is 26.5, so the area is 26.5w.

331

Thus, using length and area models, we can show that no matter what w is,

$$15.7w + 10.8w = 26.5w$$

You should choose a value of w and check that it works.

These situations offer two examples of the <u>distributive property of multiplication over addition</u>. For short, we call it the "distributive property," or "distributivity."

Distributive Property:

For any real numbers a, b, and c
$ac + bc = (a + b)c$

Examples: Simplifying using the Distributive Property

1. $4x + 62x = (4 + 62)x = 66x$

2. $-2m + -3m = (-2 + -3)m = -5m$

3. $78 \cdot \frac{2}{5} + -22 \cdot \frac{2}{5} = (78 + -22)\frac{2}{5} = 56 \cdot \frac{2}{5} = \frac{112}{5}$

4. $2c + 5d + 3c + 9d = 2c + 3c + 5d + 9d$

    $= (2 + 3)c + (5 + 9)d$

    $= 5c + 14d$

Examples: Solving equations using the Distributive Property

1. Solve: $3x + 10x = 52$

    Solution: Distributive property: $(3 + 10)x = 52$

    Simplify: $13x = 52$

    $M_{\frac{1}{13}}$: $x = 4$

332

2. Solve: $.02y + 3 + .06y \leqslant 2.44$

Solution: Assemblage property of +: $.02y + .06y + 3 \leqslant 2.44$

Distributive property: $.08y + 3 \leqslant 2.44$

A_{-3}: $.08y \leqslant -.56$

$M_{\frac{1}{.08}}$: $y \leqslant -7$

Questions covering the reading

1. According to the distributive property, $2 \cdot x + 3 \cdot x =$
   (a) $6 \cdot x$ (b) $5 \cdot x$ (c) $5 \cdot x \cdot x$ (d) $6 \cdot x \cdot x$

2. Check your answer to Question 1 by substituting -4 for x.

3. $\underbrace{a + a + \ldots + a}_{9 \text{ terms}} + \underbrace{a + a + a + \ldots + a}_{11 \text{ terms}} =$ _____.

4-9. Simplify:

4. $1.3t + 3.7t$ 5. $7y + 2y$

6. $-2\frac{1}{2}b + -3b$ 7. $\frac{1}{2}d + \frac{1}{2}d$

8. $6x + -9x$ 9. $-2m + 20m$

10. Check that $4\pi + 7\pi = 11\pi$ by using the approximation 3.14 for π.

11. (a) Simplify: $-2x + 5x$.
    (b) Check your answer by substituting 11 for x.

12. (a) Simplify: $-3m + -7m$.
    (b) Check your answer by substituting 2 for m.

13. Multiple Choice. The distributive property $(m + n)a = ma + na$ is true:
    (a) only when m and n are positive integers.
    (b) only when m and n are integers.
    (c) only when m and n are rational numbers.
    (d) when m and n are any real numbers.

333

14-27. Simplify.

14. 14a + 22a

15. $\frac{2}{3}$n + $\frac{1}{2}$n

16. $\frac{1}{2}$m + 9m + 4m

17. 4. 3B + . 3B + .4B

18. $\frac{1}{2}$m + 9m + 4

19. 4. 3B + . 3 + .4B

20. -3y + -1.4y + 2z + 3z

21. -3f + 4g + 4f + -6g

22. 3x + 5y + 2x + 9y

23. 6a + -4a + 8d + 5d

24. 9c + 6 + 3c

25. -2 + 8d + 6d

26. 2e + -6e + -14

27. 3f + 3f + 3f + 3f + 3

28-31. Give two expressions which answer each question. Thus use each situation as an example of the distributive property.

28. One day you buy 5 candy bars at 13¢ each. Another day you buy 2 candy bars at 13¢ each. How much have you spent in all?

29. One branch of a store sells m television sets at $249. 95. A second branch sells n television sets at this price. How much has been taken in?

30. Dimensions of the rectangles are given. What is the total area?

31. Dimensions of the rectangles are given. What is the total area?

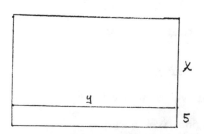

334

32-39. Solve:

32. $6x + 2x = 4$

33. $-30 > 7y + -5y$

34. $-3a + -4a = \dfrac{1}{7}$

35. $2.3B + -3.5B = .24$

36. $3 + 4c + -5c \leqslant -20$

37. $-6d - 11 + -5d > 0$

38. $2y + 3y + 8y < -39$

39. $-500x + 28 + 620x = 628$

Questions testing understanding of the reading

1-6. Using the distributive property, each of the following problems can be done easily in your head. Calculators are not allowed. Simplify:

1. $97 \cdot 5 + 3 \cdot 5$

2. $8\dfrac{1}{2} \cdot 3 + 1\dfrac{1}{2} \cdot 3$

3. $\dfrac{1}{4} \cdot \dfrac{11}{17} + \dfrac{3}{4} \cdot \dfrac{11}{17}$

4. $\dfrac{1}{2} \cdot -947 + \dfrac{1}{2} \cdot -947$

5. $15 \cdot 73 + -5 \cdot 73$

6. $80 \cdot 62 + 10 \cdot 62 + 10 \cdot 62$

7-8. In 1970, approximately 113.7 lbs of beef, 2.9 lbs of veal, 3.3 lbs of lamb or mutton, and 66.4 lbs of pork were consumed per person in the U.S. Together beef, veal, lamb, mutton, and pork are called "meats."

7. For a city of 100,000 people, approximately how many pounds of meats were consumed?

8. For each n people, approximately how many pounds of meats were consumed?

9-10. Three windows are pictured. Each has the shape of a rectangle. Dimensions are in meters.

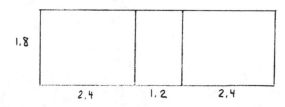

9. How much glass is in the two left windows?

10. How much glass is in all three windows?

335

11-12. In a singing group there are 4 girls and 5 boys. For singing duets, how many boy-girl pairs are possible:

11. if 2 more girls join the group.

12. if g more girls join the group.

13-14. Two slices of bacon have 5 grams of protein. Two scrambled eggs have 14 grams of protein. How many grams will you have taken in if you eat these:

13. for 6 days.   14. for n days.

15-18. Simplify:

15. $3a + 4b + 5c + 7b + 12c + 6a + 9c + 5$

16. $-11x + 2y + 3y + -2z + 2x - 4 + -5x$

17. $6m + 6m + 6m + 5n + 5n + 5n + 2p + 2p + 2p$

18. $\frac{1}{3}p + \frac{2}{3}q + \frac{4}{3} + \frac{5}{3}p + \frac{7}{3}q + \frac{8}{3}$

19-22. Simplify, then solve.

19. $9 + -9m + -5m = 2$

20. $-2z - 2 + -2z \geq 2$

C 21. $2.54x + 11.38x + 31.24 = 77.64$

22. $6y + 5 + 4y + 3 + 2y + 1 < 0$

Lesson 6

The Distributive Property and Subtraction

Here is a question which can be answered in 2 ways.

Question: In h hours, how much <u>more</u> energy is used
              by a 150-watt bulb than by a 100-watt bulb?

Answer 1:   In h hours, a 150-watt bulb uses 150h watt-
              hours of energy. In h hours, a 100h watt-
              bulb uses 100h watt-hours of energy. So the
              larger bulb uses <u>150h - 100h</u> more watt-hours .

Answer 2:   Each hour the bigger bulb uses 50 more watts
              of power. So in h hours, <u>50h</u> more watt-
              hours are used.

Of course the answers are equal. And the distributive property
verifies this.

$$150h - 100h \quad = \quad 150h + {}^-100h \qquad \text{(Def. of subtraction)}$$
$$= \quad (150 + {}^-100)\,h \qquad \text{(Distributivity)}$$
$$=\ 50h$$

Notice that by using the definition of subtraction, the subtraction
can be converted to an addition. Then the distributive property
can be applied. This happens so often that no one wants to go
through all three steps every time. It is better to learn the following property, technically called the "distributive property of

337

multiplication over subtraction." Our name is a little shorter.

Distributive Property
  with Subtraction:

> For any real numbers a, b, and c:
>
> $$ac - bc = (a-b)c$$

Examples: Distributive   Property with Subtraction

1. $15x - 8x = (15 - 8)x$

$$= 7x$$

2.   Sometimes you may still want to convert to addition.

$2A - 8A - 5B - 6B = 2A + {}^-8A + {}^-5B + {}^-6B$ (Def. of Subt.)

$$= (2 + {}^-8)A + (-5 + {}^-6)B$$ (Dist. Prop.)

$$= {}^-6A + {}^-11B$$ (Arithmetic)

$$= {}^-6A - 11B$$ (Def. of Subt.)

3.   Solve: $3a - 5a = 7$

$$-2a = 7$$

$$a = -\frac{7}{2}$$

In Example 2, the final answer was given as $\underline{{}^-6A - 11B}$, not as $\underline{{}^-6A + {}^-11B}$. This was to avoid having the symbols + and - next to each other.   Though you may want to change everything to addition to do a problem, in answers you should have only one sign between numbers or variables.

338

In answers, change   a + ⁻b to a - b.

change   a - ⁻b to a + b.

Otherwise the two signs become confused.

Questions covering the reading

1-2.   Answer each question in 2 different ways.

1.   In h hours, how much more energy is used by a 75-watt bulb than a 60-watt bulb?

2.   Each day, Melissa works 8 hours while Dana works 7.5 hours. In d days, how many more hours will Melissa have worked?

3.   What is the distributive property with subtraction?

4.   What is the technical name for the property in Question 3?

5-16.   Simplify.

5. $13x - 4x$ 6. $4z - 5z$

7. $9y - 9.5y$ 8. $⁻6w - 8w$

9. $\frac{1}{2}t - \frac{1}{3}t$ 10. $7u - 9u - 8u$

11. $3x - 4y - 2x + 3y$ 12. $⁻11m + 4m - 15m$

13. $15a - 15a$ 14. $⁻6m + 4n - 6m$

15. $-4B - 4B - 4B$ 16. $⁻3x + 10x - 3x - 10x$

17.   Why is an answer like $5x - 3z$ preferred to $5x + ⁻3z$?

18.   What answer would be preferred to $16t - ⁻6u$?

19-22.   Solve and check.

19. $9x - 5x = 84$ 20. $120 = 3y - 8y$

21. $-3T - 4T + 6T < 35$ 22. $-2A - 2A \geqslant 4$

339

1. Give a reason for each step.
   (a) $mx - nx = mx + {}^-nx$
   (b) $\qquad = (m + {}^-n)x$
   (c) $\qquad = (m - n)x$

2. A student thought that $^-5y - 5y$ simplifies to 0. Show that the student is wrong.

3. In t hours going at 45 km per hour, a car travels $45t$ km.
   (a) In the same amount of time, at 50 km per hour, how far will a second car travel?
   (b) In this time, how much farther will the second car go?

4. Plan A: Sell m tickets at \$3.00 and n tickets at \$2.50.
   Plan B: Sell m tickets at \$2.75 and n tickets at \$2.00.
   How much more money will plan A bring in?

5. <u>Multiple Choice</u>. Which cannot be simplified?
   (a) $2x - 2x$ (b) $^-2x - 2x$
   (c) $2x + {}^-2x$ (d) $2 - 2x$

6. <u>Multiple Choice</u>. Which can be simplified?
   (a) $9 - 8y$ (b) $9 + 8y$
   (c) $^-9 - 9y$ (d) $^-9y - 9y$

7.

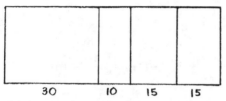

30   10   15   15

d

Dimensions are in meters.

A person wants 10,000 sq m of space for a store. Walking by a set of vacant stores, it is possible to measure their widths. They are given in the above picture. But the depth is not known. How deep must these stores be to give the required area?

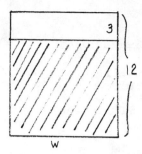

8. If the area of the shaded
   rectangle is 180, find w.

9-10. Solve and check.

9. $8m - 2 - 10m = -100$ 10. $-14n - 10n + 61 = 87$

Chapter Summary

The lesson headings point out the three related ideas which run through this chapter. They are the translation of a wide variety of situations into linear expressions, the solution of certain kinds of linear sentences, and the distributive property.

Among the situations which lead to linear expressions are some telephone rates, filling orange crates, weights or populations with constant gains or losses, calories from eating, and some movements of storms.

The sentences $ax + b = c$ and $ax + b < c$ are each solved in the same way. Add $-b$ to both sides. This results in a sentence of a type which was solved in the previous chapter.

The distributive property is assumed true for all real numbers a, b, and c:

$$(a + b)c = ac + bc$$

A variant of this property involves subtraction.

$$(a - b)c = ac - bc$$

These properties enable many expressions to be simplified.

342

CHAPTER 8

LINEAR EXPRESSIONS AND DISTRIBUTIVITY, Part II

Lesson 1

Models, Postulates, and Theorems

Many people think that the properties of numbers are uncon-
nected. These people are always confused because there seem to
be so many properties. If a person thinks every property is unrelated to others, then it is harder to learn them.

There are lots of properties. Even

$$4x + 3x = 7x$$

is a property! So the properties must be organized. Some are
more important than others. In this book, the named properties
are of three types: models, postulates, and theorems.

Models are properties which connect operations with the
real world or with uses of numbers. When they involve the real
world, models are not always exact. (For example, there is an
area model for multiplication, but you cannot calculate a real
area exactly.) But the models tell us why the operations are important. And they suggest precise properties of the operations.

When a property is suggested by a model, you can use the
model to check whether it is true. But there are only two sure

343

ways to find out whether a property will always work.

(1) Assume the property is true.

(2) Show that the property follows from
other more basic properties already
assumed true.

Those properties which are assumed true are called <u>postulates</u>.
Over the past two hundred years, the following list of postulates
has been developed. All other properties of addition or multiplication of real numbers follow from these postulates. You have
now studied all the properties of addition and multiplication which
are postulates.

Postulates of Addition and Multiplication of Real Numbers

Let a, b, and c be any real numbers. Then:

Closure Properties:

$a + b$ is a real number. ab is a real number.

Assemblage Properties:

$$a + b = b + a \qquad ab = ba$$

$$a + (b + c) = (a + b) + c \qquad a(bc) = (ab)c$$

Existence Properties:

There is a unique real number 0 with a + 0 = a.

There is a unique real number 1 with a · 1 = a.

There is a unique real number -a with a + -a = 0.

When a ≠ 0, there is a unique real number $\frac{1}{a}$ with a·$\frac{1}{a}$ = 1.

Distributive Property:

$$ac + bc = (a + b)c$$

Sentence-Solving Properties:

If a = b, then a + c = b + c.

If a = b, then ac = bc.

If a < b, then a + c < b + c.

If a < b and c > 0, then ac < bc.

If a < b and c < 0, then ac > bc.

Notice that there are a good number of postulates. But
there aren't so many that they can't be learned.

Important properties which follow from the postulates
are called <u>theorems</u>. The "distributive property with subtraction"

$$ac - bc = (a - b)c$$

is an example of a theorem. So is the oppositing property of
-1, that -1 · x = -x.

In this chapter, more theorems are discussed and applied.
These theorems are found by combining the distributive property with the other postulates.

<u>Questions covering the reading</u>

1. What is a model?

2. What are the two ways to find out whether a property will
always work?

3. What is a postulate?   4. What is a theorem?

5. Give an example of a postulate.

6. Give an example of a theorem.

7. Give an example of a model.

8. Choose the most general word: model, property, postulate,
theorem.

9-14. Let x, y, and z be any real numbers. Name the postulate
which guarantees the truth of each statement.

9. $x + y = y + x$ 10. $yx + zx = (y + z)x$

11. If $x < y$, then $x + z < y + z$.

12. $z(xy) = (zx)y$

13. $y + z$ is a real number.

14. There is a number 0 with $x + 0 = x$.

15-20. Use variables to describe each property.

15. closure property of multiplication

16. additive identity property

17. associative property of addition

18. commutativity of multiplication

19. addition property of equations

20. multiplication property of inequalities

21. One postulate could be called the "Property of Reciprocals." What postulate is that?

Questions testing understanding of the reading

1-10. Each given statement follows from one of the postulates listed in this lesson. Name the postulate. (If the statement were more important, it would be called a theorem.)

1. $-3 + 462 = 462 + {}^-3$

2. $153 \cdot \dfrac{1}{153} = 1$

3. If $3x = 9$, then $x = 3$.

4. $2m + {}^-9m = {}^-7m$

5. $9(x + 3) = (x + 3) \cdot 9$

6. If $9y + 6 < 10$, then $9y < 4$.

7. $5y = 5y + 0$

8. $5 \cdot 3x = (5 \cdot 3)x$

9. $m \cdot 1 = m$

10. $\pi + 4$ is a real number

Question for discussion and exploration

1. The word "postulate" has many synonyms. (a) What is a synonym? (b) Look in a dictionary to determine which of the following is not a synonym of "postulate": axiom, assumption, conclusion, hypothesis.

Lesson 2

Distributivity and the Multiplicative Identity Property

The multiplicative identity property says that for any number a, $1 \cdot a = a$. When this property is combined with distributivity, you can simplify some expressions which you have studied before.

1. An item has price p. If the price is raised 20%, what is the new price?

   Solution: You already know that the new price is $\underline{p + .20p}$. Now you can simplify this expression.

   $$p + .20p = 1 \cdot p + .20p$$
   $$= (1 + .20)p$$
   $$= 1.20p$$

   The new price is 1.20 times the old price.

2. A bookcase selling for \$249.95 is lowered 15% in price. What is the new price?

   Solution: The answer is $249.95 - .15(249.95)$. But this can be simplified before calculating.

   $$249.95 - .15(249.95) = 1 \cdot 249.95 - .15(249.95)$$
   $$= .85(249.95)$$

   That is, the new price is 85% of the old price.

   In general: $p - .15p = .85p$

These kinds of simplifications help to solve sentences.

3. An auto dealer quotes you a price of \$3500 for a new car and says: "This is only 10% above my cost." What is the dealer's cost?

Solution: Let c be the dealer's cost.

$$\text{Then} \quad c + .10c = 3500$$
$$1 \cdot c + .10c = 3500$$
$$(1 + .10)c = 3500$$
$$1.10c = 3500$$
$$c = \frac{3500}{1.10} \approx 3181.01$$

The dealer's cost is near \$3180.

4. A \$50,000 estate is to be split up among three children and a sister of the person who died. If each child gets the same amount and the sister gets half this much, how much should the sister expect to receive?

Solution: Let c be what each child gets. The sister then gets $\frac{1}{2}c$.

$$\text{So} \quad c + c + c + \frac{1}{2}c = 50,000$$
$$3\frac{1}{2}c = 50,000$$
$$c = \frac{50,000}{3.5} \approx 14,286$$

The sister should expect $\frac{1}{2}c$, or about \$7143.

1-10. Simplify.

1. $7a + a$ 2. $4t - t$

3. $v - 9v$ 4. $w + 2w$

5. $p + .06p$ 6. $m - .34m$

7. $x - 2x - x$ 8. $-14y - 7y + y$

9. $q - .3q$ 10. $r - r - r - r - 2r$

11-16. Give an expression which answers each question. If the expression can be simplified, do so.

11. The price of a CB radio is originally g. The price is raised 10%. What is the new price?

12. A calculator's price is reduced 30%. If the original price was x, what is the new price?

13. You put an amount P in a savings account which gives 6% interest in a year. How much will you have at the end of one year?

14. The population of the world is presently X and growing at the rate of 2% a year. What will be the population one year from now?

15. Some taxicab companies allow their drivers to keep 55% of all fares taken in. The rest goes to the company. If a driver takes in P dollars in fares, how much does the company get?

16. If you have a total income of t dollars and the government gets $\frac{3}{10}$ of your income in taxes and social security, how much do you have left?

17-20. Solve.

17. $p + .05p = 21$ 18. $B - 11B = 62$

19. $d + 4d + 9 < -8$ 20. $g - .30g + 4 \geq 109$

350

Questions applying the reading

1-8. (a) Find a sentence which can be used to answer the given question. (b) Answer the question.

C 1.   After a 20% discount, the price of a television set is $279.96. What was the original price of the set?

C 2.   The normal markup in shoe stores is about 38% of the dealer's price. If you buy a pair of shoes for $16.95, what did it cost the shoe dealer?

C 3.   You are a retailer and want an item to cost $10 <u>including</u> tax for a sale. (This is sometimes done during sales to make it easier to "ring up" the charges. Then more customers can be treated.) If the sales tax is 5%, how much should you charge for the item without the tax?

C 4.   Suppose a retailer collects 4% sales tax. On a particular day the retailer forgot to keep sales tax separated from the rest of the receipts. Total receipts for the day were $458.64. (a) About how much should the retailer pay in tax? (b) Why is the answer to part (a) only an estimate?

5.   A man dies and leaves his estate to his wife and three children. The total value of the estate was $29460. If each child got one-half as much as his wife gets, how much did each child get? How much did his wife get?

6.   A woman dies and leaves her estate to her husband and two children. The total value of the estate was about $42,000. If each child gets one-third of what her husband gets, how much should each child get? How much should her husband get?

C 7.   One author of a book is to receive twice as much as a second author. If the total royalties from sales of the book are $1,549.26, how much should each author receive?

8.   A home team is to get 5 times as much from the gate as a visiting team. If the total receipts are $1572, how much should each team receive?

9-10. The answers to these problems are not as obvious as you may think.

9.   Suppose that prices go up 10% in the next year and then down 10% the following year. How will prices that second year compare with present prices?

10. Due to financial problems, a city lowers the pay of its workers by 5%. Then, after negotiations, it raises the pay 5%. Is this fair to the workers?

Question for discussion

C 1.   In the plumbing business, it is common to give a discount of "6 10's and a 20." This means that a price is lowered 10%, then that price is lowered 10%, then that third price is lowered 10%, etc., six times, and finally the price is lowered 20%. On an original price of P dollars, what will be the final price after "6 10's and a 20"?

Lesson 3

Zero, Multiplication, and Division

You know how to multiply by zero. (You've already learned the theorem!) So play a game in this lesson. Act as if you didn't know what $5 \cdot 0$ or $0 \cdot 173.2$ is. How could you figure it out?

There are two ways to figure out a multiplication problem you don't know. One way is to use a <u>model</u> for multiplication. For example, here is the area model.

a ▢   The area of the rectangle is <u>a • b</u>.

b

352

Now what happens when b = 0?

a |

The area of the figure is <u>a · 0</u>.

But the area is 0.

So a · 0 = 0.

A second way to figure out a multiplication problem is to use theorems or postulates. The only postulate which involves 0 is the additive identity property. For any number a,   0 + a = a.
But we want to consider multiplication. So we multiply both sides
by any number x.   (0 + a)x = ax
By distributivity: 0 · x + ax = ax
Now A_{-ax}:   0 · x = 0
In this way, from the postulates, we get a property already known to you.

Zero Multiple
Theorem:

For any real number x,

$$0 \cdot x = 0$$

All equations you solved earlier had exactly one solution.
The zero multiple theorem shows that certain equations do not have just one solution.

353

$$0 \cdot x = 0 \qquad\qquad 0 \cdot x = 3$$

has infinitely many   has no solution, for
solutions, for any   the left side is
real number works.   always 0, never 3.

Multiplying by zero causes all kinds of problems when solving

sentences. Here is a different sort of example.

$$2y = 9 \qquad \text{only solution } 4.5$$

$M_0: \qquad 0 \cdot 2y = 0 \cdot 9$

$$0y = 0 \qquad \text{now any number works!}$$

If you started with $2y < 9$, you would get $0y < 0$ and no number

would work.

> Never multiply both sides
>
> of a sentence by 0. You
>
> may gain or lose solutions.

Division by zero is a little more complicated than multipli-

cation. Two cases must be considered, division by zero and divi-

sion into zero.

Division by zero: Since $0 \cdot x = 0$ always, it is never true

that $0 \cdot x = 1$. So 0 has no reciprocal. Recall that dividing by

b is the same as multiplying by the reciprocal of b. Since 0

has no reciprocal, you cannot divide by it. Division by 0 is impos-

sible.

You can check this idea using the rate model of division.

$$\frac{x}{y} \text{ meters per second} = \frac{x \text{ meters}}{y \text{ seconds}}$$

Try to think what $\frac{10 \text{ meters}}{0 \text{ seconds}}$ would mean. It can't be done. You cannot travel 10 meters in 0 seconds just as it is meaningless to think of $\frac{10}{0}$.

Division into zero: You can travel 0 meters in 10 seconds. That is a rate of $\frac{0}{10}$ meters per second, which is 0 meters per second. Thus the rate model suggests that $\frac{0}{b} = 0$, a property easily verified by the zero multiple theorem.

$$\frac{0}{b} = 0 \cdot \frac{1}{b} \qquad \text{(Def. of division)}$$

$$= 0 \qquad \text{(Zero multiple theorem)}$$

Everything with division gives rise to some result about fractions. The above arguments show:

| The denominator of a fraction cannot be zero. | When the numerator of a fraction is zero, the value of the fraction is zero. |

355

1. What is the Zero Multiple Theorem?

2. Suppose you do not know how to figure out a multiplication or division problem. What two things can you use to help you?

3. A "rectangle" has length 5 cm and width 0 cm.
   (a) Draw a picture of such a "rectangle."
   (b) What is the area of the "rectangle?"

4. What property or definition is used in each step?
   (a) For any number x, $\quad\quad 0 + a = a$
   (b) $\quad\quad\quad\quad\quad\quad\quad\quad (0 + a)x = ax$
   (c) $\quad\quad\quad\quad\quad\quad\quad\quad 0 \cdot x + ax = ax$
   (d) $\quad\quad\quad\quad\quad\quad\quad\quad 0 \cdot x = ax + {}^-ax$
   (e) $\quad\quad\quad\quad\quad\quad\quad\quad 0 \cdot x = 0$

5-6. What does each situation have to do with division?

5. You cannot go 150 km in 0 seconds.

6. If you go 0 km in 2.5 minutes, you have traveled at a rate of 0 km per minute.

7-8. Choose the meaningless expression.

7. $3 \div 0$ or $0 \div 3$ $\quad\quad\quad\quad$ 8. $\dfrac{0}{x + 5}$ or $\dfrac{x + 3}{0}$

9-12. What value can x **not** have in each expression?

9. $\dfrac{4}{x}$ $\quad\quad\quad$ 10. $\dfrac{0}{x}$ $\quad\quad\quad$ 11. $\dfrac{x}{14}$ $\quad\quad\quad$ 12. $\dfrac{3}{x - 2}$

13. Give an example to show that it is unwise to multiply both sides of an equation by zero.

14. Give an example to show that it is unwise to multiply both sides of an inequality by zero.

15-23. Give all solutions to each sentence.

15. $x \cdot 0 = 0$ $\quad\quad\quad$ 16. $0y = 50$ $\quad\quad\quad$ 17. $0 + z = 75$

18. $A \cdot 0 = {}^-2$ $\quad\quad\quad$ 19. $\dfrac{1}{x} = 0$ $\quad\quad\quad$ 20. $\dfrac{0}{y} = 0$

21. $\dfrac{1}{3} = 0 \cdot B$ $\quad\quad\quad$ 22. $0 = m \cdot 0$ $\quad\quad\quad$ 23. $0 = 3n$

<u>Questions testing understanding of the reading</u>

1-10. Each situation leads to a multiplication or division problem which involves zero. (a) What is the problem? (b) Answer the question.

1. Suppose you have $4\frac{1}{2}$ drinks and none of them has any alcohol. How much alcohol will you have drunk?

2. Profit from a store is to be split up between the 2 owners. But there is no profit. How much did each owner get?

3. A rich man's estate is to be divided among his children. But he has no children. How much did each child get?

4. If the left blank can be filled in 6 ways and the right blank cannot be filled, then _____ _____ in how many ways can the two blanks be filled?

5. A 100-watt bulb is kept off for 24 hours. How many watt-hours of electrical power have been used?

6. You have spent no money in 4 hours in a store. How fast have you been spending?

7. You have spent $25 in 0 minutes. How fast have you been spending?

8. You gain 3 g in weight in 0 seconds. What is your rate of weight gain?

9. You have no gain in weight for 4 days. What is your rate of weight gain?

10. A person is paid nothing per hour for doing 10 hours work. How much is the total payment?

11-16. Give all solutions to each sentence.

11. $2x - 2x = 0$

12. $4y + {}^-4y = 4$

13. $0 \leqslant \frac{1}{2}z$

14. $1.63 < w + 0$

15. $A \cdot 0 > 10$

16. $4.1B \geqslant 0$

17-20. What value can the variable <u>not</u> have?

17. $\dfrac{1}{x+3}$ 18. $\dfrac{2+y}{4+y}$ 19. $\dfrac{z+1}{3z}$ 20. $\dfrac{2x-4}{x+8}$

21-22. Solve.

21. $\dfrac{2+x}{5}=0$ 22. $\dfrac{4y-200}{36}=0$

Question for discussion

1. In an earlier lesson, you learned the oppositing property of $^-1$. $^-1 \cdot x = {^-x}$. Show that this property is a theorem by giving the reason for each step. (There are lots of steps because nothing is missing. This problem is easiest if you do it one line at a time.)

  (a) $^-1 \cdot x = {^-1} \cdot x + 0$
  (b) $= {^-1} \cdot x + (x + {^-x})$
  (c) $= ({^-1} \cdot x + x) + {^-x}$
  (d) $= ({^-1} \cdot x + 1 \cdot x) + {^-x}$
  (e) $= ({^-1} + 1)x + {^-x}$
  (f) $= 0x + {^-x}$
  (g) $= 0 + {^-x}$
  (h) $= {^-x}$

Lesson 4

Distributivity and Commutativity

Situation 1: In a particular year, a grandparent decides to give a birthday present of $5 to each grandchild. If this person has s grandsons and d granddaughters, how much will be given?

Answer 1: The grandsons receive 5s dollars.
The granddaughters will get 5d dollars.
Total given: 5s + 5d dollars.

Answer 2: There are <u>s + d</u> grandchildren. Each gets $5.
Total given: 5(s + d) dollars.

358

The situation suggests that 5s + 5d and 5(s + d) are equal. The 5 could be replaced by any real number. So the following property seems reasonable. This property is labelled a theorem because it follows from the postulates of distributivity and commutativity.

Theorem:
(Commutativity
variation of
distributivity)

For any real numbers a, b, and c,

$$c(a + b) = ca + cb$$

Argument (to show that the theorem follows from earlier properties):

$c(a + b) = (a + b)c$	Commutativity of $\cdot$
$= ac + bc$	Distributivity
$= ca + cb$	Commutativity of $\cdot$

Examples: 1. $3(x + 4)$ 2. $-8(2 - y)$

$= 3 \cdot x + 3 \cdot 4$ $= -8(2 + -y)$

$= 3x + 12$ $= -8 \cdot 2 + -8 \cdot -y$

                                  $= -16 + 8y$

Suppose a telephone call costs 22¢ for the first 3 minutes and 6¢ for each additional minute. Then an m‑minute call will cost .22 + .06(m - 3). You might wonder how long can you call for under $1.00 at these rates. To find out, solve:

$$.22 + .06(m - 3) < 1.00$$

$$.22 + .06m - .18 < 1.00 \Bigg\lbrace \begin{array}{l} \text{Applying the theorem} \\ \text{leads to a sentence} \\ \text{you know how to} \\ \text{solve.} \end{array}$$

Simplify: $\qquad .04 + .06m < 1.00$

$A_{-.04}:$ $\qquad\qquad .06m < .96$

$M_{\frac{1}{.06}}:$ $\qquad\qquad m < 16$

So you must talk for less than 16 minutes. This means you should talk for at most 15 minutes; otherwise you will be charged for 16.

A special case of the theorem arises in the following way:

Situation 2: You begin a day with B sheets of paper.
In your first class you use u sheets.
In your second class you use v sheets.
How many sheets are left?

Answer 1: You have used u + v sheets. So you must have <u>B - (u + v)</u> sheets left.

Answer 2: After the first class you had B - u sheets left. Then you used v sheets. So you must have <u>B - u - v</u> sheets left.

This situation suggests that $B - (u + v) = B - u - v.$

In general, it means that $-(u + v) = -u - v.$ This property can be shown to follow from other properties.

<u>Opposite of a Sum Theorem:</u>

> For any real numbers a and b,
>
> $-(a + b) = -a - b$

360

Argument (to show that this property follows from earlier prop-

   erties):

$$-(a + b) = -1 \cdot (a + b) \qquad \text{Oppositing property of } -1$$

$$= -1 \cdot a + -1 \cdot b \qquad \text{Earlier theorem (p. 359)}$$

$$= -a + -b \qquad \text{Oppositing property of } -1$$

$$= -a - b \qquad \text{Def. of subtraction}$$

Examples: 4. $-(2t + u) = -2t - u$

5. $-(8 - m) = -8 - -m = -8 + m$

6.   Solve: $-(2x - 5) = 40$

Solution: $-2x + 5 = 40$

$$-2x = 35$$

$$x = -\frac{35}{2} = -17.5$$

Questions covering the reading

1-4.   Give two expressions which answer each question.

1.   The 4 children in the Bell family are each given n dollars
     by their uncle and m dollars by their aunt during a holiday
     visit. Altogether how much was given to the children?

2.   A dealer buys 11 TV sets for c dollars each and sells them
     for s dollars each. What is the dealer's total profit?

3.   One day 20 rows of a scarf are knitted, each row containing
     120 stitches. The next day 12 rows are knitted, each row
     containing 120 stitches. How many stitches were done in
     the two days?

4. Suppose that in one bank account you have x_1 dollars of savings at 5% interest per year. In a second bank you have x_2 dollars of savings earning 5% interest. How much interest should you expect to get in a year?

5. What is a theorem?

6. According to a theorem of this lesson, $c(a + b) =$ _____.

7. Give the assumed property which tells why each step can be made.
   (a) $t(3 + u) = (3 + u)t$
   (b) $= 3t + ut$

8-17. Find an equal expression which has no parentheses.

8. $8(a + b)$ 9. $14(2c + d)$

10. $-3(2u + v)$ 11. $-1(-3 + x)$

12. $\frac{1}{2}(6W - 10)$ 13. $\frac{2}{3}(4y - 9)$

14. $3(a + b - c)$ 15. $m(-2n + 4)$

16. $d(-2 - e)$ 17. $-\frac{7}{8}(-16a - b)$

18. The opposite of a number equals _____ times that number.

19. If $-462 = q \cdot 462$, then $q =$ _____.

20. If $-(a + b) = x(a + b)$, then $x =$ _____.

21-24. Give two expressions which answer each question.

21. You begin the day with $5.26. You spend $1.25 on lunch and $3.50 for a concert ticket. How much do you have left?

22. You begin the day with B dollars, spend L dollars on lunch, and c dollars for a concert ticket. How much do you have left?

362

23. What is the area of the shaded part of this 10 by 10 square?

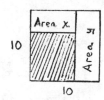

24. The entire segment has length L. What is the length which remains after parts with lengths c and d are cut off?

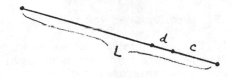

25-32. Find an equal expression which has no parentheses.

25. $-(2 + 6x)$

26. $-(t - 4)$

27. $-(3v - 1)$

28. $4 - 2(x + y)$

29. $-6 - 8(\frac{1}{2} + 2m)$

30. $-100 - 80(-3q - r)$

31. $a - (b - c)$

32. $x - (y + z)$

33-40. Solve:

33. $3(2x - 11) = -8$

34. $5(2 + y) = 2$

35. $1 \geq .20(z - 5)$

36. $12 < 3(4A + 6)$

37. $1 - (B - 2) = 6$

38. $11 - (3C + 4) = 17$

39. $8 + 2(4 + 6D) < 0$

40. $7 - 2(y - 12) > 4$

Questions testing understanding of the reading

1-4. Do these problems in your head. No calculators are allowed.

1. $9 \cdot 8 + 9 \cdot 12$

2. $\frac{1}{2} \cdot 49 + \frac{1}{2} \cdot -49$

3. $71 \cdot 105 - 71 \cdot 5$

4. $13 \cdot 11 + 13 \cdot -1$

5. $3.2 \cdot 9.93 + 3.2 \cdot .07$

6. $5683 \cdot 4723 - 5683 \cdot 4722$

7. $-60(\frac{1}{6} + \frac{1}{5})$

8. $100(\frac{1}{4} + \frac{1}{5})$

9-10. If a long-distance call costs $.30 for the first 3 minutes and $.12 for each additional minute, then a t-minute phone call will cost <u>.30 + .12(t - 3)</u>.

9. How long can you talk for $3?

10. How long can you talk for less than $5?

11-12. In summer, the weather bureau puts out a number called the "temperature humidity index" or THI. It is calculated from Fahrenheit temperatures as follows:

$$THI = .4(W + D) + 15 \qquad \begin{array}{l} W = \text{wet bulb thermometer reading} \\ D = \text{dry bulb thermometer reading} \end{array}$$

11. A TV weatherman announces a (dry bulb) temperature of 80 and a THI of 73. What must have been the wet bulb thermometer reading?

12. A majority of people are uncomfortable if THI > 75. If the dry bulb reading is 90, what must be the wet bulb reading in order for most people to be uncomfortable?

13-20. Solve these more difficult sentences.

13. $2(n + 4) - 3n = 8$

14. $9(m - 2) + 2(m + 4) = 12$

15. $100(y + 2) + 10y = 420$

16. $-(x - 3) - (3 - x) = 4$

17. $-(2 - 5A) - 2A > 14$

18. $3 - 3(B + 6) \le 0$

19. $10 - \frac{1}{2}(6C - 10) < 5$

20. $2D - 2(4 - D) = 720$

Lesson 5

Distributivity, Division, and Fractions

Two cakes are to be split among n people. One has volume v, the other has volume w. How much will each person get?

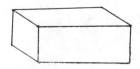

Cake with volume v   Cake with volume w

Solution 1:   Each person will get $\frac{v}{n}$ from the left cake,

$\frac{w}{n}$ from the right cake. Total received by each

person: $\frac{v}{n} + \frac{w}{n}$

Solution 2:   The total volume is $v + w$. Each person gets

$\frac{1}{n}$ of that, so gets $\frac{v+w}{n}$.

Of course the answers are equal. That is, $\frac{v}{n} + \frac{w}{n} = \frac{v+w}{n}$.
This is the way you have added fractions when they have the same
denominator. The distributive property helps to show that this
is a legal way to add fractions.

$$\frac{a}{c} + \frac{b}{c} = a \cdot \frac{1}{c} + b \cdot \frac{1}{c} \qquad \text{(Def. of division)}$$

$$= (a + b)\frac{1}{c} \qquad \text{(Distributivity)}$$

$$= \frac{a+b}{c} \qquad \text{(Def. of division)}$$

Adding
Fractions
Theorem:

> For any real numbers a, b, and c with $c \neq 0$,
>
> $$\frac{a}{c} + \frac{b}{c} = \frac{a+b}{c}$$

Examples: Adding Fractions with the Same Denominator

1. $\frac{2x}{5} + \frac{4x+3}{5} = \frac{2x+(4x+3)}{5}$

$$= \frac{6x+3}{5}$$

2. $\frac{y}{2} - \frac{3+y}{2} = \frac{y + {}^-(3+y)}{2}$

$$= \frac{y-3-y}{2}$$

$$= \frac{-3}{2}$$

3. $\frac{2}{w} + \frac{3}{w} = \frac{2+3}{w} = \frac{5}{w}$

365

Examples: Breaking up Fractions with Additions in the Numerator

1. $\dfrac{n+5}{5} = \dfrac{n}{5} + \dfrac{5}{5}$ 2. $\dfrac{-40x-60}{-10} = \dfrac{-40x}{-10} + \dfrac{-60}{-10}$

   $= \dfrac{n}{5} + 1$ $= 4x + 6$

Questions covering the reading

1. Give reasons for each step.

   (a) $\dfrac{x}{3} + \dfrac{2y}{3} = x \cdot \dfrac{1}{3} + 2y \cdot \dfrac{1}{3}$

   (b) $= (x + 2y) \cdot \dfrac{1}{3}$

   (c) $= \dfrac{x + 2y}{3}$

2. What is the Adding Fractions Theorem?

3. On what property is the Adding Fractions Theorem based?

4-6. Write the answer to each question in two different ways.

4. Two cakes, one with p pieces and one with q pieces, are to be split up among n people. How much will each person get?

5. A brother and sister want to pool money to buy a typewriter. Each will contribute half of what they earn. The brother earns b dollars. The sister earns s dollars. How much will they contribute together?

6. If you weigh w kg on Earth, you would weigh about $\dfrac{w}{6}$ kg on the moon. What would be the total weight on the moon of (1) an astronaut weighing w kg on Earth and (2) the astronaut's gear which weighs 60 kg on Earth?

7-14. Simplify:

7. $\dfrac{3}{10} + \dfrac{11}{10}$ 8. $\dfrac{7}{x} + \dfrac{-7}{x}$

366

9. $\dfrac{3x}{y} + \dfrac{x}{y}$

10. $-\dfrac{a}{2} + \dfrac{5a}{2}$

11. $\dfrac{999}{4000} + \dfrac{1}{4000}$

12. $\dfrac{3n}{4} + \dfrac{2+n}{4} - \dfrac{2}{4}$

13. $\dfrac{4B}{5} - \dfrac{B}{5}$

14. $-\dfrac{9m}{t} - \dfrac{9m}{t}$

15-20. Break up each fraction.

15. $\dfrac{4x - 6}{2}$

16. $\dfrac{9y + 12x + 15}{3}$

17. $\dfrac{m + 5 + 4m}{5}$

18. $\dfrac{-10a - 22a}{-a}$

19. $\dfrac{7xy + y}{y}$

20. $\dfrac{-2 + 2x}{2}$

21. How can you check your answer to a question like #19 above?

Questions testing understanding of the reading

1-3. These questions are related.

1. (a) What is the mean of two numbers a and b?
   (b) True or False? You can calculate the mean of two numbers by taking half of each number and adding the halves.

2. (a) What is the mean of three numbers a, b, and c?
   (b) True or False? You can calculate the mean of three numbers by taking half of each number and adding the halves.

3. Develop a shortcut to calculate the mean of each set of 4 numbers. (Hint: Question 1 or Question 2 may help.)
   (a) 40. 44, 48, 32
   (b) -1.2, - 2.8, -4.8, -3.2
   (c) 3636, 4812, 2080, 52

4-7. Simplify one of the fractions.   Then add.

4. $\dfrac{2x}{6} + \dfrac{5}{3}$

5. $\dfrac{y - 3}{2} + 4\dfrac{1}{2}$

6. $\dfrac{9z}{6} + \dfrac{z}{2}$

7. $\dfrac{100}{4z} + \dfrac{2}{z}$

8-15. Simplify.

8. $\dfrac{8}{y-2} + \dfrac{3}{y-2}$

9. $\dfrac{x+3}{x} - \dfrac{3}{x}$

10. $\dfrac{7000}{692} - \dfrac{80}{692}$

11. $\dfrac{2a}{6} + \dfrac{10a}{6} - \dfrac{3b}{6} - \dfrac{9b}{6}$

12. $\dfrac{3}{t} - \dfrac{t-3}{t}$

13. $\dfrac{8x}{5} - \dfrac{3+8x}{5}$

14. $\dfrac{2y-8}{4} - \dfrac{2y}{4}$

15. $\dfrac{13}{a} - \dfrac{6-a}{a}$

16-19. Break up each fraction.

16. $\dfrac{15x+40}{20}$

17. $\dfrac{42a-12}{6}$

18. $\dfrac{63a+7ab+7}{7a}$

19. $\dfrac{-4x-8y}{-2}$

Lesson 6

Decision-Making Using Sentences

If you have a telephone in many places in the U.S., you have
a choice of rates. The following monthly rates are similar to
those found in Chicago for people who make a lot of calls.

Choice 1
Base rate of $11.25 for
200 calls plus .0523 for
each call over 200.

Choice 2
$24.50 for an unlimited
number of calls.

How do you decide which plan is better? The _easiest_ way is to set up a sentence. Let n be the number of calls which are made in a month. (We assume n > 200. Otherwise Choice 1 is the obvious choice.) How much will it cost to make n calls?

cost using Choice 1	cost using Choice 2
11.25 + .0525(n - 200)	24.50

When is Choice 1 better? Clearly it is better when the cost using Choice 1 is less than the cost using Choice 2. So the idea is to solve:

$$11.25 + .0525n - 10.50 < 24.50$$
$$.0525n + .75 < 24.50$$
$$.0525n < 23.75$$
$$n < \frac{23.75}{.0525} = 452.38\ldots$$

Choice 1 is better if n is 452 or less. That is, if you make 452 or fewer phone calls a month, Choice 1 is better. Otherwise Choice 2 is cheaper.

> Solving a sentence can
> often help make a decision.

Sometimes both choices need variables to describe them. Then the sentence to be solved has variables on both sides. Here are two examples.

Example 1: One machine can sort 300 cards a minute. A second machine can sort 400 cards a minute but takes 5 more minutes to warm up. When should you use the second machine?

Solution: Let t be the number of minutes Machine 1 is in use. Then Machine 2 is in use for $t - 5$ minutes. We assume $t > 5$. (Otherwise it is obvious that Machine 1 should be used.)

cards sorted in t minutes using Machine 1	cards sorted in t minutes using Machine 2
$300t$	$400(t - 5)$

Machine 2 will sort more cards when

$$300t < 400(t - 5)$$

In the next lesson, you will study how to solve this sentence. Its solution is

$$20 < t$$

So if a job takes more than 20 minutes (during which over 6000 cards can be sorted), Machine 2 should be used.

370

Example 2:   In one school district, teachers with a master's
               degree are paid $9000 plus $500 for each year's
               teaching experience. In a second district, teach-
               ers are paid $9750 plus $350 for each year's
               experience. (These are rather typical salaries
               for 1976 in some parts of the country.) How
               many years experience are needed for a teacher
               to make more for working in the first district?

Solution: Let E be the number of years teaching exper-
               ience. Then:

               salary in   salary in
               District 1   District 2

               9000 + 500E   9750 + 350E

               You can answer the question that was asked by

               solving

                    9000 + 500E > 9750 + 350E

Questions covering the reading

1-3. Suppose that the cost of an n-minute long distance call under
rate plan 1 is .10 + .20n. Under rate plan 2 the cost is .80 +
.15(n - 3). What sentence should you solve to find out:

1.   when the call would cost less under the first plan.

2.   when the call would cost less under the second plan.

3.   when calls would cost the same under the two plans.

371

4. One machine can sort 500 cards a minute. A second machine can sort 700 cards a minute but takes 2 minutes longer to warm up. In m minutes of use:
   (a) How many cards can be sorted by the first machine?
   (b) How many cards can be sorted by the second machine?
   (c) What sentence can you solve to find out when the first machine should be used?

5. In one area, teachers' salaries are $8500 a year plus $600 for each year's experience. In a second area, the salaries are $9200 a year plus $450 for each year's experience. After E years experience:
   (a) How much would you make in the first area?
   (b) How much would you make in the second area?
   (c) What sentence can you solve to find out when teachers make more in the first area?

Questions testing understanding of the reading

1-4. Here are 4 rent-a-car plans. These rates are not unusual. Gas is never included.

Company A:	$15.95 a day plus 14¢ a mile
Company B:	$12.95 a day plus 15¢ a mile
Company C:	$9.95 a day plus 17¢ a mile
Company D:	$10.95 a day plus 19¢ a mile

1. (a) Give the cost of driving n miles with Company A.
   (b) Give the cost of driving n miles with Company B.
   (c) What sentence would you solve to find out when Company A is cheaper?

2. (a) Give the cost of driving m miles with Company C.
   (b) Give the cost of driving m miles with Company D.
   (c) What sentence would you solve to find out when Company C is cheaper?

3. (a) Give the cost of driving d miles with Company A.
   (b) Give the cost of driving d miles with Company D.
   (c) What sentence would you solve to find out when Company A is cheaper?

4. (a) Give the cost of driving x miles with Company B.
   (b) Give the cost of driving x miles with Company C.
   (c) What sentence would you solve to find out when Company C is cheaper?

Lesson 7

Solving $ax + b \leq cx + d$

This kind of sentence can be converted in one step to a kind you already know how to solve. Here are some examples.

1. Solve: $\qquad\qquad 50x + 120 = 30x - 200$

    The idea is to find an equivalent sentence with all variables on the same side. First add $-30x$ to each side.

    A_{-30x}: $\qquad -30x + 50x + 120 = -30x + 30x - 200$

    Simplify: $\qquad 20x + 120 = -200$

    How solve as before.

    A_{-120}: $\qquad\qquad 20x = -320$

    $M_{\frac{1}{20}}$: $\qquad\qquad\quad x = -16$

2. Solve: $\qquad\qquad 2 - 3y > y + 4$

    First add $3y$ to each side. (You could add $-y$ and get the same final answer.)

    A_{3y}: $\qquad 2 - 3y + 3y > y + 4 + 3y$

    Simplify: $\qquad\qquad\quad 2 > 4y + 4$

    A_{-4}: $\qquad\qquad\quad -2 > 4y$

    $M_{\frac{1}{4}}$: $\qquad\qquad\quad -\frac{1}{2} > y$

3.   Example 1 in Lesson 6 (p. 370) dealt with card sorting machines. The question led to this inequality.

$$300t < 400(t - 5)$$

Distributivity: $300t < 400t - 2000$

A_{-400t}: $-100t < -2000$

$M_{-\frac{1}{100}}$: $t > 20$

This answer means that more cards are sorted on Machine 2 when the machines are operating over 20 minutes.

The work in solving sentences of this kind can be summarized by the following algorithm.

Algorithm 5:   To solve a sentence of the form $ax + b \leq cx + d$ for x add $-cx$ to each side. Then solve the resulting sentence.

Questions covering the reading

1-4.   (a) What could you add to each side of the sentence to help solve it? (b) Solve the sentence.

1. $2x + 4 = 9x - 12$ 2. $-8y + 2 < -3y + 4$

3. $\frac{1}{2}n \geq \frac{1}{3}n + 6$ 4. $14 - 2x \leq 11 + 8x$

374

5-12. Solve.

5. $20(x + 4) = 16(x + 10)$ 6. $-2a + 3a + 7 < 4a - 29$

7. $18 + y = -3y + 6$ 8. $12B - B = 100B$

9. $-6 - 4z > 2z + 9$ 10. $100(c - 3) \geq 200c$

11. $3n + 6 \leq 2(n + 3)$ 12. $5x + 12 - 3x = -2 - 8x + 6$

13-20. Each of the following sentences arises from a situation given in Lesson 6. (a) Solve the sentence. (b) Apply the solution to answer the question asked about the situation. You will need to look back at the page where the original question was asked.

13. $9000 + 500E > 9750 + 350E$ (Example 2, p. 371 .)

14. $600E + 8500 > 450E + 9200$ (QCR 5c, p. 372.)

15. $15.95 + .14n < 12.95 + .15n$ (QTU 1c, p. 372.)

16. $9.95 + .17x < 10.95 + .19x$ (QTU 2c, p. 372.)

17. $.10 + .20n < .80 + .15(n - 3)$ (QCR 1, p. 371.)

18. $.10 + .20n > .80 + .15(n - 3)$ (QCR 2, p. 371.)

19. $.10 + .20n = .80 + .15(n - 3)$ (QCR 3, p. 371.)

20. $500m > 700(m - 2)$ (QCR 4c, p. 372.)

Questions testing understanding of the reading

1-2. Different banks offer different rate plans for writing checks. Some plans are better for those who write only a few checks a month. Other plans are better for those who write many checks.

1. Bank A charges $2.00 a month plus $.05 a check. Bank B charges $.10 a check.
   (a) How much does it cost to write c checks a month in bank A?
   (b) How much does it cost to write c checks a month in bank B?
   (c) When is it cheaper to have a checking account in bank A?

2. Bank A charges $1.50 a month plus $.08 a check. Bank B charges $2.50 a month plus $.05 a check.
   (a) How much does it cost to write n checks a month in bank A?
   (b) How much does it cost to write n checks a month in bank B?
   (c) When is it cheaper to have a checking account in bank A?

3-6. Solve for x.

3. $ax = bx + c$

4. $2x + a = 3x + 4a$

5. $3x + b < c - 9x$

6. $nx + b = nx + c$

C 7-10. Examine these world records in free-style swimming.

Distance	Sex	as of 1954	as of 1974	Change per year
100m	M	54.8 sec.	51.2 sec.	-.18 sec.
	F	64.6 sec.	58.1 sec.	-.32 sec.
200m	M	123.9 sec.	112.8 sec.	-.55 sec.
	F	141.7 sec.	123.6 sec.	-.93 sec.
400m	M	266.7 sec.	238.2 sec.	-1.42 sec.
	F	300.1 sec.	257.3 sec.	-2.14 sec.
1500m	M	1099.0 sec.	931.8 sec.	-8.36 sec.
	F	1377.0 sec.	1014.2 sec.	-18.14 sec.

In each case notice that the women's world record is getting closer to the men's world record. In all distances but the 100 meters the present women's record is faster than the men's record was in 1954. If the change per year remains the same:

7. (a) What would be the men's 100 meter record n years after 1974?
    (b) What would be the women's 100 meter record n years after 1974?
    (c) When would the records be the same?

8. (a) What would be the men's 200 meter record t years after 1974?
    (b) What would be the women's 200 meter record t years after 1974?
    (c) When would the records be the same?

9. (a) What would be the men's 400 meter record x years after 1974?
   (b) What would be the women's 400 meter record x years after 1974?
   (c) When would the records be the same?
   (d) What would be the world record at the time of your answer to part (c)?

10. (a) What would be the men's 1500 meter record y years after 1974?
    (b) What would be the women's 1500 meter record y years after 1974?
    (c) When would the records be the same?
    (d) What would be the world record at the time of your answer to part (c)?

Lesson 8

Situations Which Always or Never Happen

Which job would you take?

Job 1	Job 2
starting salary $2.60/hour;	starting salary $2.50/hour;
every 3 months increases	every 3 months increases
$.10/hr.	$.10/hr.

Of course, the answer is obvious. Job 1 will always pay better than Job 2. But what happens when this is solved mathematically? Let n = number of 3-month periods worked.

pay in Job 1	pay in Job 2
$2.60 + .10n$	$2.50 + .10n$

377

When is the pay in Job 1 better than the pay in Job 2? The corresponding mathematical problem is to solve

$$2.60 + .10n \; > \; 2.50 + .10n$$

$A_{-.10n}:$ $2.60 + .10n + -.10n \; > \; 2.50 + .10n - .10n$

Simplify: $2.60 \; > \; 2.50$ The variable seems to
have disappeared!

What has happened is that we have

$$2.60 + 0 \cdot n \; > \; 2.50$$

$$0 \cdot n \; > \; -.10$$

and this sentence is always true. The sentence $\underline{2.60 \; > \; 2.50}$ is
always true. Write as the answer: <u>n may be any real number.</u>

> If in solving a sentence you get a sentence
> which is always true, the original sentence
> is always true.

When does Job 1 pay less than Job 2? To answer this, you
could solve:

$$2.60 + .10n \; < \; 2.50 + .10n$$

$A_{-.10n}:$ $2.60 \; < \; 2.50$

It is never true that 2.60 is less than 2.50. So Job 1 never pays
less than Job 2, something which was obvious from the pay rates.

> If in solving a sentence, you get a
> sentence which is never true, the
> original sentence is never true.

Examples:

1. Solve: $5 + 3x = 3(x - 2)$

    Solution: $5 + 3x = 3x - 6$

    A_{-3x}: $5 = -6$

    Write: never true, so original sentence has no solution

2. Solve: $8(2y + 5) < 16y + 60$

    Solution: $16y + 40 < 16y + 60$

                $40 < 60$

    Write: always true, so original sentence is true for every
                possible value of y

As a final example, look at a problem about populations.

Here are two cities. When will their population be the same?

City 1	City 2
Present population 23,000	Present population 23,000
Growing 1,000 a year	Growing 1,200 a year

In t years:

Population of City 1	Population of City 2
$23000 + 1000t$	$23000 + 1200t$

The populations will be the same when

$$23000 + 1000t = 23000 + 1200t$$

Solving:

$$A_{-1000t}: \qquad 23000 = 23000 + 200t$$

$$A_{-23000}: \qquad 0 = 200t$$

$$M_{\frac{1}{200}}: \qquad 0 = t$$

The t = 0 means their populations are the same <u>now</u>.

> Zero being a solution <u>is different</u>
>
> <u>from</u> having no solution at all.

Questions covering the reading

1. (a) Add ⁻2x to both sides of this sentence. What sentence results?

   $$2x + 10 < 2x + 8$$

   (b) What should you write to describe the solutions to this sentence?

2. (a) Add 5y to both sides of this sentence. What sentence results?

   $$^{-}5y + 9 = 3 - 5y + 6$$

   (b) What should you write to describe the solutions to this sentence?

3-10. Solve:

3. $2(2y - 5) \leqslant 4y + 6$ 4. $3x + 5 = 5 + 3x$

5. $^{-}2m = 3 - 2m$ 6. $6c = 3(2 + 2c)$

7. $2A - 10A > 4(1 - 2A)$ 8. $9y - y < 9$

9. $2B + B = 5B$ 10. $18v + 6 > 9(4 + 2v)$

11. The population of City 1 is about 200,000 and growing at about 5,000 people a year. City 2 has a population of about 185,000 and is growing at about 5000 people a year. In t years:
   (a) What will be the population of City 1?
   (b) What will be the population of City 2?
   (c) When will their populations be the same?

12. Answer Question 11 if City 2 has a population of 200,000 and is growing at a rate of 4000 people a year.

Questions testing understanding of the reading

1. Here are two of the rent-a-car plans first given in Lesson 6.

   Company C: $9.95 a day plus .17 a mile
   Company D: $10.95 a day plus .19 a mile

   (a) Give the cost of driving d miles with Company C.
   (b) Give the cost of driving d miles with Company D.
   (c) What sentence would you solve to find out when Company D is cheaper?
   (d) Solve the sentence and interpret your answer.

2. Refer to Question 1.

   (a) What sentence would you solve to find out when Company C is cheaper?
   (b) Solve the sentence and interpret your answer.

Chapter Summary

As in the previous chapter, three related ideas run through this chapter. They are the distributive property and its many variations, the translation of a wide variety of situations into linear expressions and sentences, and the solution of linear sentences.

Models have suggested many properties of the operations. Those which are assumed true are called <u>postulates</u>. The distributive property is one of a small number of statements normally taken as postulates of addition and multiplication of real numbers. A list of these postulates is on page 345.

Many theorems can be derived by combining the distributive property with definitions or other postulates:

with the multiplicative identity property $a + ca = (1 + c)a$

with commutativity: $c(a + b) = ca + cb$

with the definition of division: $\dfrac{a + b}{c} = \dfrac{a}{c} + \dfrac{b}{c}$

with the additive identity property: $a \cdot 0 = 0$

All linear sentences are equivalent to sentences of the type

$$ax + b \lesseqgtr cx + d$$

Adding $-cx$ to both sides results in an equation of the form

$$ex + b \leq d$$

which can be solved by the methods of the last chapter. When
$e = 0$, a sentence results which has no solution or all real numbers as solutions.

The solution of these linear sentences was applied to telephone and rent-a-car rates, weights or populations with constant
gains or losses, salaries and costs and discounts of all kinds.

INDEX

Absolute value 159, 160
Adding Fractions Theorem 364
Addition
  models 120, 122, 130
  properties 133, 134, 139, 345
Addition Property
  of Equations 280
  of Inequalities 291
Additive Identity 138
Al-Khowarizmi 273
Algebra 68
Algorithm 267
  solving ax = b 269, 270
  solving a + x = b 281
  solving ax + b = c 321
  solving ax + b < c 326
  solving ax + b ≤ cx + d 373
Analysis 68
Arabic numerals 47
Area 89
Area Model
  for Multiplication 187, 194, 250
Assemblage Property
  of Addition 134
  of Multiplication 195
Associative Property
  of Addition 134
  of Multiplication 194
Average 153

Bar graph 16

Calculus 252
Centimeter 54, 55
Circle graph 21
Coding 2
Coefficient 268
Commutative Property
  of Addition 134
  of Multiplication 193, 359
Comparison 20, 21, 235
Computer science 68
Conversion
  length 55
  temperature 92
  weight or mass 55

Counting 1, 95, 12
Cutting-off Model
  of Subtraction 147

Decimals 47, 59
  infinite 49, 59
  repeating 49
Descartes, Rene 112
Deviation 161, 162
Difference 74
Diophantus 84
Directed Distance Model
  of Subtraction 148
Distance 165
Distributive Property
  with Subtraction 338
Division 218, 244, 245
  by zero 354
  models 178, 217, 237, 244
Equations 263, 280
Equivalent sentences 263
ERA 278
Event 226, 227
Experiment 34
Extremes 276

Factor 238
Fair coin, die 222
Formula 89
Fraction
  addition 364
  multiplication 246
Fraction Simplification Property
  239

Geometry 68
Gram 54
Graphs
  bar 16
  circle 21
  number line 14, 43, 100, 287

Hamilton, William Rowan 134
Harriot, Thomas 73
Hindus 47